SPRINGER TRACTS IN MODERN PHYSICS

Ergebnisse
der exakten Natur-
wissenschaften

Volume **63**

Editor: G. Höhler

Editorial Board: P. Falk-Vairant S. Flügge J. Hamilton
F. Hund H. Lehmann E. A. Niekisch W. Paul

Springer-Verlag Berlin Heidelberg GmbH 1972

Manuscripts for publication should be adressed to:

G. Höhler, Institut für Theoretische Kernphysik der Universität, 75 Karlsruhe 1, Postfach 6380

Proofs and all correspondence concerning papers in the process of publication should be addressed to:

E. A. Niekisch, Kernforschungsanlage Jülich, Institut für Technische Physik, 517 Jülich, Postfach 365

ISBN 978-3-662-15576-9   ISBN 978-3-540-37461-9 (eBook)

DOI 10.1007/978-3-540-37461-9

Photon-Hadron Interactions II

International Summer Institute in Theoretical Physics DESY, July 12—24, 1971

Other lectures given at this meeting will be found in Volume 62 of "Springer Tracts in Modern Physics"

Contents

---

* This contribution was not among the
lectures presented at the International Summer
Institute

High Energy Photoproduction
of Pseudoscalar Mesons

J. Frøyland

Contents

Introduction

Low energy photoproduction may be studied in an almost model-independent way by means of phase shifts, but high energy photoproduction has to be studied in terms of models simply because phase shift analysis becomes impossible at sufficiently high energies.

Since good data first became available a large number of models has been proposed to "explain" the data. Unfortunately, none of these distinguish themselves as appearing to be definitely more correct than the others, although it seems that the models that have done best so far are of the Regge pole plus Regge cut types. The cuts being either absorptive or more phenomenological. However, also these models have met difficulties [1] and should not be taken too seriously.

Recognizing these facts the following discussion will be to a large extent qualitative, and we will try to extract consequences of the models that can be understood without calculations and that are common to as many models as possible. Below we list some of the models that have been tested on photoproduction. Some of them exist in several versions and some are incomplete in the sense that they try to describe only a limited amount of data. Together with the list are given some references, but this list is not claimed to be complete.

1) Strong absorption [2] (Michigan).
2) Mixed model [3].
3) Weak absorption [3, 4] (Argonne).
4) Other absorptive models [5–11].
5) Multiple scattering [12].
6) Pseudo-model [13].
7) Phenomenological cuts [14–17].
8) Complex poles [18].
9) Conspiracy models [19–22].
10) Electric Born term [23].
11) Veneziano model [24–28].
12) Fixed poles [29].
13) Vector dominance model.

The order in which the models appear does not imply any judgement on their relative correctness.

Most of what will be said in the following sections is fairly easily adapted to models 1)–6). However, model 3) in its original form is unable to fit the data [3].

Complex poles will be neglected essentially because of their lack of predictive power. The conspiracy models and the electric Born term models seem unlikely to be correct for reasons to be given later.

There have also been some attempts to use the Veneziano model. In principle, it probably makes more sense to study the Veneziano four-point function in πN photoproduction than in πN scattering because of the absence of the Pomeron. The problem of introducing spin in the Veneziano model still seems hard and is possibly connected to the fact that the models, that have been tried, suffer from having parity doublets and a pion conspirator. Satellite terms are needed to connect the forward and backward directions [28]. This is also considered undesirable.

The linearity of the unitarity relations for photoproduction implies that fixed poles are not forbidden. The fact that (at least for charged photoproduction) $d\sigma/dt \sim s^{-2}$ has lent support to the assumption that fixed poles rather than moving (Regge) poles are dominant. Essentially, the arguments that will be given later against the electric Born term model also apply to fixed poles.

The vector dominance model will hardly be mentioned because it is the topic of another lecture at the school.

General Considerations

The basic ingredients in all models trying to describe two-body or quasi two-body processes is the assumed dominance of t channel exchanges[1]. Therefore, the first step in any conventional analysis of two-body scattering processes is to establish the possible exchanges.

Table 1. Table of pseudoscalar photoproduction processes. The crosses indicate that the experiment has been done. The line between the measured target asymmetry for $\gamma p \to K^+ \Sigma^0$ and $p \to K^+ \Lambda$ indicates that Λ and Σ^0 were not distinguished from another

	$d\sigma/dt$	Pol. beam	Pol. target
$\gamma p \to \pi^+ n$	×	×	×
$\gamma n \to \pi^- p$	×	×	
$\gamma p \to \pi^0 p$	×	×	×
$\gamma n \to \pi^0 n$	×		
$\gamma p \to \eta p$	×		
$\gamma n \to \eta n$			
$\gamma p \to \pi^+ \Delta^0$	×	×	
$\gamma n \to \pi^+ \Delta^-$	×		
$\gamma p \to \pi^- \Delta^{++}$	×	×	
$\gamma n \to \pi^- \Delta^+$	×		
$\gamma p \to \pi^0 \Delta^+$			
$\gamma n \to \pi^0 \Delta^0$			
$\gamma p \to K^+ \Lambda$	×		×
$\gamma n \to K^0 \Lambda$			
$\gamma p \to K^+ \Sigma^0$	×		×
$\gamma n \to K^+ \Sigma^-$	×		
$\gamma p \to K^0 \Sigma^+$	×		
$\gamma n \to K^0 \Sigma^0$			
$\gamma p \to K^+ Y^0$	×		
$\gamma n \to K^+ Y^-$	×		
$\gamma p \to K^0 Y^+$			
$\gamma n \to K^0 Y^0$			

A list of the processes so far studied experimentally is given in Table 1. All processes listed (above) below the double line proceed via (non) strange exchanges.

As is seen from the Table the most extensive experimental studies have been done on photoproduction of pions off nucleons. The same statement is no doubt true for the theoretical efforts. Most of the features

---

[1] Of course, the duality picture tells us that we gain extra information if s channel resonances are included in the analysis, but at the moment we will ignore this.

of these processes cannot teach us very much without some labour on the spinology of the problem. It is, therefore, interesting to note that one can get essential pieces of information from the data on $\gamma N \to \pi \varDelta$ and $\gamma N \to K \varSigma$ without elaborate models.

If we believe in any type of absorbed Regge model, it implies that each amplitude can be written symbolically as $R + RP$, where R represents the contribution from a pure Regge pole and RP represents the Regge-Pomeron cut induced by the absorption. Allowing the Pomeron to be exchanged in the initial and final states does not change the quantum numbers of the t channel exchanges with the exception of parity as we shall see later.

In first order perturbation theory we have to consider simple one-particle exchange diagrams. These will give rise to factors $\langle \pi | j_\mu(0) | B \rangle$ where B represents the exchanged particle. The electromagnetic current may be written as a sum of an isovector and an isoscalar current[2]

$$j_\mu(x) = j_\mu^{(s)}(x) + j_\mu^{(v)}(x) . \tag{1}$$

Under G parity

$$\begin{aligned} j_\mu^{(s)} &\to -j_\mu^{(s)} \\ j_\mu^{(v)} &\to j_\mu^{(v)} \end{aligned} \tag{2}$$

and from this follows that $I^G = 0^+, 2^+$ are not allowed as quantum numbers for the exchanged particles.

If we look at the simple one-particle exchange diagram for $\gamma N \to \pi \varDelta$ we see that only $I = 1, 2$ may be exchanged, but no particle with $I = 2$ is known, so, if we have an $R + RP$ model $I = 1$. Looking at the Clebsch-Gordan coefficients, we find

$$d\sigma(\gamma p \to \pi^- \varDelta^{++})/dt \sim 3 \sum_{i=1}^{8} |M_i^{(s)} - M_i^{(v)}|^2$$

$$d\sigma(\gamma n \to \pi^+ \varDelta^-)/dt \sim 3 \sum_{i=1}^{8} |M_i^{(s)} + M_i^{(v)}|^2$$

$$d\sigma(\gamma n \to \pi^- \varDelta^+)/dt \sim \sum_{i=1}^{8} |M_i^{(s)} - M_i^{(v)}|^2 \tag{3}$$

$$d\sigma(\gamma p \to \pi^+ \varDelta^0)/dt \sim \sum_{i=1}^{8} |M_i^{(s)} + M_i^{(v)}|^2$$

where M_i are the eight-independent t channel amplitudes and the indices s and v denote photon isospin.

---

[2] Here we have ignored the possibility that the current may also have an isotensor part. This point has been discussed by Professor Donnachie in his lectures.

This means that with the assumptions of t channel dominance and no $I = 2$ exchange it follows that

$$d\sigma(\gamma p \to \pi^- \Delta^{++})/dt = 3d\sigma(\gamma n \to \pi^- \Delta^+)/dt \qquad (4)$$

and

$$d\sigma(\gamma n \to \pi^+ \Delta^-)/dt = 3d\sigma(\gamma p \to \pi^+ \Delta^0)/dt . \qquad (5)$$

Disregarding deuteron effects

$$R_{\pi^+} \equiv \frac{d\sigma}{dt}(\gamma d \to \pi^+ \Delta N)\bigg/\frac{d\sigma}{dt}(\gamma p \to \pi^+ \Delta^0) = 4 , \qquad (6)$$

$$R_{\pi^-} \equiv \frac{d\sigma}{dt}(\gamma d \to \pi^- \Delta N)\bigg/\frac{d\sigma}{dt}(\gamma p \to \pi^- \Delta^{++}) = \frac{4}{3} . \qquad (7)$$

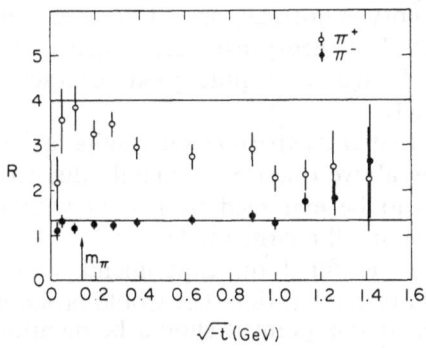

Fig. 1. The ratios $(\gamma d \to \pi^+ \Delta N)/(\gamma p \to \pi^+ \Delta^0)$ (open circles) and $(\gamma d \to \pi^- \Delta N)/(\gamma p \to \pi^- \Delta^{++})$ (closed circles) versus $(-t)^{\frac{1}{2}}$ at 16 GeV/c. From Ref. [30]

Correcting for deuteron effects brings down the ratio R_{π^+} a bit but not sufficiently to be in agreement with the experiment [30]. See Fig. 1.

Requiring the absence of $I = \frac{3}{2}$ exchanges in the $\gamma N \to K \Sigma$ processes gives

$$\frac{d\sigma}{dt}(\gamma d \to K^+ \Sigma N)\bigg/\frac{d\sigma}{dt}(\gamma p \to K^+ \Sigma^0) = 3 \qquad (8)$$

which again is a bit bigger than the experimental values [31].

Doing the analysis with the "exotic" $I = 2$ and $I = \frac{3}{2}$ exchanges included, it can be seen that a contribution of about 10–20% of exotic exchanges in the amplitudes is sufficient to account for the data. The principal explanations for the observed discrepancies are:

1. Regge-Regge cuts;
2. s channel effects;
3. exotic particle exchanges;

4. tensor current;
5. deuteron effects;
6. background effects.

The first three explanations accept that there is an effect, while the two last ones cast some doubt on the results. The iso-tensor current explanation is impossible because the Clebsch-Gordan coefficients are so that it cannot be distinguished from the iso-scalar part. If any one of the first three explanations is right it is a bit disturbing to note that sum rules following from the requirement of absence of exotics have recently been tested [32] and found always to be satisfied within errors for purely hadronic processes above 4–5 GeV. If one believes in the vector dominance model there should be no reason why the sum rules should not be satisfied also in photoproduction processes.

The Glauber theory is normally used to extract neutron target data from deuteron data. The theory has been tested in a large number of cases and seems to be true to a quite good accuracy. This makes also explanation 5 unlikely.

The possibility that the experimental results (at least for $\gamma N \rightarrow \pi \Delta$) may not be entirely above question is mainly due to the fact that the cross-sections have to be extracted from data with background that may not be the same in all charge modes.

It would be very useful if the experiments were done at several energies. If RR cuts are in operation they would probably show up in the energy dependence. At this point it should be mentioned that a backward peak has been observed in $K^- p \rightarrow K^- p$ scattering [33]. This process can proceed only by "exotic" exchanges in the u channel. A theoretical calculation using RR cuts [34] is in rough agreement with the data, but it is not clear if the predicted energy dependence will be correct.

Before we proceed to look in greater detail at the other processes, we should bear in mind the lessons we have learned from the preceding, namely that there may be contributions of the order of 10–20% in the amplitudes that may not be explained by any $R + RP$ model.

Absorption Type Models

Concepts like Regge-Pomeron cuts or Regge-Regge cuts are very important in our understanding (or lack of understanding) of two-body or quasi two-body processes, so a short review of some basic ideas may be useful.

Qualitatively speaking, the procedure is to Reggeize in the t channel and then cross to the s channel where the absorption is performed. For the time being let us assume that the Reggeization and crossing have been done. The s channel Regge pole amplitudes may then be written as $M^R_{\lambda' \mu'; \lambda \mu}$ where λ, μ and λ', μ' are the helicities of the incoming and outgoing particles respectively. For later use we make the following definitions:

$$n \equiv |(\lambda' - \mu') - (\lambda - \mu)| \,, \tag{9}$$

$$x \equiv |\lambda - \lambda'| + |\mu - \mu'| - n \,. \tag{10}$$

In order to get a feeling of the significance of these variables let us look at the partial wave expansion of the amplitudes

$$M_{\lambda' \mu'; \lambda \mu} = \sum_J (2J + 1)\, a^J_{\lambda' \mu'; \lambda \mu}\, d^J_{\lambda' - \mu'; \lambda - \mu}(\theta) \,. \tag{11}$$

The d^J functions contain factors $\sin^n(\theta/2)$ and, therefore, $M(\theta = 0) = 0$ if $n \neq 0$. Let us also neglect the problems of unequal mass kinematics so that $t = 0$ corresponds to the forward direction. Then, if the Regge poles factorize in the s channel (which we shall make plausible later) we see that there is a factor $(\sqrt{-t})^{|\mu - \mu'|}$ associated with the upper vertex of a simple one-Reggeon-exchange diagram, and a similar factor $(\sqrt{-t})^{|\lambda - \lambda'|}$ associated with the lower vertex which give altogether a factor $(\sqrt{-t})^{n+x}$. Consequently, since $x \geq 0$ a Regge pole amplitude will be non-zero in the forward direction only when $n = 0$ and $x = 0$, while kinematics alone demands only $n = 0$. Already from this one sees immediately that all t channel pole contributions have to vanish at $t = 0$ in pseudoscalar meson photoproduction simply because

$$n + x = |\lambda_\gamma - \lambda_\pi| + |\lambda_{N_i} - \lambda_{N_f}| = 1 + |\lambda_{N_i} - \lambda_{N_f}| > 0 \,.$$

Let us see what happens if we include absorption. According to the Sopkovich prescription [35], each partial wave amplitude of the Regge pole amplitude M^R_l is transformed in the following way

$$M^{\text{abs}}_l = (S^{\text{el}}_l)^{\frac{1}{2}}\, M^R_l (S^{\text{el}}_l)^{\frac{1}{2}} \,. \tag{12}$$

S^{el}_l is the S matrix element for elastic scattering in the state with angular momentum l. We have not distinguished between initial and final state elastic scattering and all helicity indices have been suppressed. Putting

$$S^{\text{el}} = 1 + i\varrho M^{\text{el}} \tag{13}$$

where ϱ is a simple kinematical factor we get

$$M^{\text{abs}}_l = M^R_l + i M^R_l M^{\text{el}}_l \varrho \,. \tag{14}$$

The last term now represents the partial waves of the RP cut. Since the elastic scattering is predominantly imaginary and positive, it follows from Eq. (14) that the interference between the cut and the pole terms will be essentially destructive.

It is customary to make the impact parameter representation which means

$$l \to qr - \tfrac{1}{2}$$

$$\sum_{l=0}^{\infty} \to \int_{0}^{\infty} q \, dr \qquad (15)$$

and

$$d^{J}_{\lambda\mu}(\theta) \approx J_n(r\sqrt{-t}). \qquad (16)$$

Parametrizing

$$M^{el} \sim q^2 \exp(At) \qquad (17)$$

one gets

$$M_l^{el} \sim A^{-1} \exp(-l^2/4q^2 A) \qquad (18)$$

which is a Gaussian in l. This means that the lowest partial waves are more absorbed than the higher partial waves.

Following Ref. [2] we will be a bit more precise and assume

$$M^{el}_{\lambda'\mu';\lambda\mu} = 4q^2 \sigma_{tot}(i + \varrho) \, e^{At/2} \delta_{\lambda\lambda'} \delta_{\mu\mu'} \qquad (19)$$

where $\varrho = \mathrm{Re}\, M(0)/\mathrm{Im}\, M(0)$. Using the impact parameter representation and doing a somewhat lengthy calculation one ends up with a formula for the cut in the form of a convolution integral

$$M^{c}_{\lambda'\mu';\lambda\mu} = -(4\pi)^{-1} \sigma_{tot}(1 - i\varrho) \, e^{At/2}$$
$$\cdot \frac{1}{2} \int_{-\infty}^{0} dt' \cdot e^{At'/2} I_n(A\sqrt{tt'}) \, M^{R}_{\lambda'\mu';\lambda\mu}(t') \qquad (20)$$

where

$$I_n(z) = (-i)^n J_n(iz) \qquad (21)$$

is real for real z.

Let us define

$$F_n(z) \equiv I_n(z)/z^n \qquad (22)$$

then $F_n(0)$ is finite and different from 0. From this it is easy to see some of the most characteristic features of the absorptive Regge cut models. Remember that at some fixed energy we may write

$$M^{R}(t) = (-t)^{\frac{1}{2}(n+x)} f(t) \qquad (23)$$

where $f(t)$ contains all residue functions, choosing mechanism factors, signature factors and propagators. Inserting in Eq. (20) the result is

$$M^{c}_{n} \sim e^{At/2}(\sqrt{-t})^n \frac{1}{2} \int_{-\infty}^{0} dt' \cdot (-t')^{n+x/2} e^{At'/2} F_n(A\sqrt{tt'}). \qquad (24)$$

The factor $(\sqrt{-t})^n$ in front of the integral is exactly the required kinematical factor. $F_n(z)$ is a slowly varying function with z. Since $e^{At'/2} f(t')$ decreases very rapidly with increasing $-t'$ it is clear that the essential part of the integral will come from the small $-t'$ region and, therefore, depend very crucially on the value of $n + x/2$. Clearly, the effect of absorption will be big in $n = 0$ amplitudes, less in $n = 1$ amplitudes and small in $n = 2$ amplitudes. It is also obvious that the effect will be especially dramatic in amplitudes of the type $(n, x) = (0, x > 0)$ which will be non-vanishing at $t = 0$. (Remember that all pole contributions to such amplitudes have to vanish at $t = 0$.)

The Michigan group [2] argues that in the formula for M^c one has not taken into account the fact that there may be inelastic intermediate states induced by the same exchanges. Because of the energy phase relation such states will add coherently. For this reason the replacement $M^c \to \lambda M^c$ is made with $\lambda \geq 1$ real to take care of the inelastic intermediate states.

When λ gets so big that the lowest partial waves change sign, one characterizes the situation by saying that the partial waves have been over-absorbed. If there is no over-absorption in an $(n, x) = (0, x \geq 0)$ amplitude there will be, in general, two dips in the modulus of the amplitude simply because the cut will dominate at very small $-t$ and at large $-t$ and the pole will dominate in between. When the over-absorption is made sufficiently large the two dips collapse to one. This is, in fact, what happens in the Michigan model [2] for the π exchange in charged π photoproduction from nucleons. The large over-absorption is "necessary" to fit the data at $t = 0$ but has the secondary effect of giving a cut that dominates at large t.

In spite of the relative successes of the absorptive type models and their intuitive appeal, it should be emphasized that there is no fundamental theory behind them, and consequently, the absorption may be performed in other equally well (or badly) justified ways.

Kinematics for $\gamma N \to 0^- \frac{1}{2}^+$

For our analysis it is convenient to construct the t channel parity conserving helicity amplitudes [36]. For simplicity we shall write

$$M^{(t)\pm}_{\lambda_\gamma - \lambda_\pi; \lambda_{\bar{N}} - \lambda_N} \equiv M^{(t)\pm}_{\lambda_\gamma \lambda_\pi; \lambda_{\bar{N}} \lambda_N} \tag{25}$$

where M^+ and M^- represent parity conserving helicity amplitudes with contributions from natural $[P = (-1)^J]$ and unnatural $[P = -(-1)^J]$ parity states, respectively.

By analyzing the quantum numbers of the $\bar{N}N$ system one easily finds to which amplitudes the various poles contribute. A listing is given in Table 2. Doing the crossing [37] and keeping only terms of leading order in s we get the following results:

$$M^{(s)}_{0-\frac{1}{2};1\frac{1}{2}} \sim M^{(t)-}_{10} + M^{(t)+}_{11} \qquad (n=0)$$

$$M^{(s)}_{0\frac{1}{2};1-\frac{1}{2}} \sim M^{(t)-}_{10} - M^{(t)+}_{11} \qquad (n=2) \qquad (26)$$

$$M^{(s)}_{0\frac{1}{2};1\frac{1}{2}} \sim M^{(t)+}_{10} + M^{(t)-}_{11} \qquad (n=1)$$

$$M^{(s)}_{0-\frac{1}{2};1-\frac{1}{2}} \sim M^{(t)+}_{10} - M^{(t)-}_{11} \qquad (n=1).$$

Here some kinematic factors irrelevant for our purposes have been neglected.

Table 2. Pole contributions to the various t channel parity conserving helicity amplitudes

$(\lambda\mu)^\pm$	Isospin index $(-)$	(0)	$(+)$
$(10)^+$	A_2	ϱ	ω
$(10)^-$	π	B	—
$(11)^+$	A_2	ϱ	ω
$(11)^-$	A_1	—	—

The t channel parity conserving helicity amplitudes are supposed to factorize and from Eq. (26) it is clear that also the pole contributions to the s channel amplitudes will factorize.

From what we know about kinematics and absorptive cuts we see that in general

$$M_{0-\frac{1}{2};1\frac{1}{2}}(t=0) \neq 0 \qquad (n,x)=(0,2) \qquad (27)$$

but

$$M_{0\frac{1}{2};1-\frac{1}{2}}(t=0) = 0 \qquad (n,x)=(2,0). \qquad (28)$$

Translated into t channel language Eq. (28) reads

$$M^-_{10}(t=0) = M^+_{11}(t=0) \qquad (29)$$

which is often referred to as the conspiracy condition because it implies that at $t=0$ the natural and unnatural parity contributions are not independent of each others. The absorptive cuts automatically give non-trivial solutions to Eq. (29) and they are, therefore, said to be self-conspiring.

It is easily seen from Eq. (26) that the $\pm$ nature of the amplitudes M^+_{10} and M^-_{11} are preserved after the absorption, but this is not the case for the amplitudes M^-_{10} and M^+_{11}.

Let us establish that the asymmetry measured with a plane-polarized beam can distinguish between natural and unnatural parity exchanges.

The polarization vectors for photons of helicities ± 1 are

$$\varepsilon_{\pm 1} = 2^{-\frac{1}{2}}(\mp 1, -i, 0) \tag{30}$$

and those representing plane-polarized photons

$$\varepsilon^{\perp} = i2^{-\frac{1}{2}}(\varepsilon_{-1} + \varepsilon_{+1}) \quad \text{and} \quad \varepsilon^{\parallel} = 2^{-\frac{1}{2}}(\varepsilon_{-1} - \varepsilon_{+1}). \tag{31}$$

Defining the asymmetry as

$$\Sigma = \frac{d\sigma_{\perp} - d\sigma_{\parallel}}{d\sigma_{\perp} + d\sigma_{\parallel}} \tag{32}$$

and using Eq. (26) and the parity relations

$$M^{(s)}_{\lambda_\pi \lambda_{N'}; \lambda_\gamma \lambda_N} = -(-1)^{\lambda_\gamma - \lambda_N - \lambda_\pi + \lambda_{N'}} M^{(s)}_{-\lambda_\pi - \lambda_{N'}; -\lambda_\gamma - \lambda_N} \tag{33}$$

one obtains [38]

$$\Sigma = \frac{|M_{10}^+|^2 + |M_{11}^+|^2 - |M_{10}^-|^2 - |M_{11}^-|^2}{|M_{10}^+|^2 + |M_{11}^+|^2 + |M_{10}^-|^2 + |M_{11}^-|^2}. \tag{34}$$

In a very similar manner one may find the asymmetry with a polarized target, A, and the polarization, P, of the recoiling nucleon.

$$A = \frac{2\,\mathrm{Im}[M_{10}^+ M_{11}^{+*} + M_{10}^- M_{11}^{-*}]}{|M_{10}^+|^2 + |M_{11}^+|^2 + |M_{10}^-|^2 + |M_{11}^-|^2}, \tag{35}$$

$$P = \frac{2\,\mathrm{Im}[M_{10}^+ M_{11}^{+*} - M_{10}^- M_{11}^{-*}]}{|M_{10}^+|^2 + |M_{11}^+|^2 + |M_{10}^-|^2 + |M_{11}^-|^2}. \tag{36}$$

A glance at Table 2 tells us that only A_1 exchange may contribute to M_{11}^- for charged π photoproduction, and for π^0 photoproduction there is no candidate to contribute to M_{11}^-. We also remember that if the elastic scattering is diagonal in the helicities, then also the absorption will not generate contributions to M_{11}^-. This means that in a quite general way A_{π^0} is equal to P_{π^0}. This may also be approximately true in charged photoproduction because the A_1 and its associated Pomeron cut will not interfere with other poles or their Pomeron cuts. Consequently, the A_1 exchange will constitute some sort of common background to π^- and π^+ photoproduction which certainly has to be small because of the smallness of the π^-/π^+ ratio.

Especially, if one finds $A_{\pi^0} \neq P_{\pi^0}$ one would probably have to ascribe the difference to RR cuts, or an off-diagonal Pomeron contribution to a RP cut. However, the difference between A_{π^0} and P_{π^0} cannot be very big because A, P and Σ are not independent of each other. To see this we note that

$$A + P \sim 2\,\mathrm{Im}\,M_{10}^+ M_{11}^{+*}$$
$$A - P \sim 2\,\mathrm{Im}\,M_{10}^- M_{11}^{-*}$$
$$1 + \Sigma \sim |M_{10}^+|^2 + |M_{11}^+|^2 \tag{37}$$
$$1 - \Sigma \sim |M_{10}^-|^2 + |M_{11}^-|^2 .$$

Since

$$|M_{10}^+ \pm i M_{11}^+|^2 \geqq 0$$
$$|M_{10}^- \pm i M_{11}^-|^2 \geqq 0 \tag{38}$$

it follows that

$$|A - P| \leqq 1 - \Sigma$$
$$|A + P| \leqq 1 + \Sigma . \tag{39}$$

The data [39, 40] for Σ for $\gamma p \to \pi^0 p$ (Fig. 2) show that for small $|t|$ Σ_{π^0} is not far from 1 and thus $A \approx P$ no matter how big RR contributions

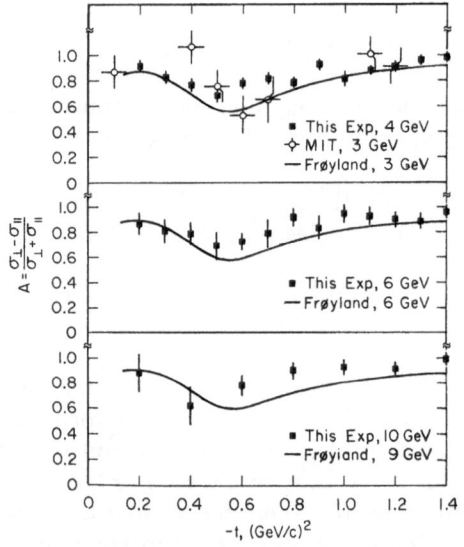

Fig. 2. Σ_{π^0} at 3, 4 and 10 GeV/c. From Ref. [39]

may be. Evidently, the best place to look for a difference between A and P is in the bottom of the dip at about $t = -0.5\ \mathrm{GeV}^2$.

In $\gamma p \to \pi^+ n$ both [41–43] Σ and [44] A have been measured and assuming $A = P$, we can make a comparison between $|A|$ and $\frac{1}{2}(1 + \Sigma)$. The comparison has been given in Fig. 3.

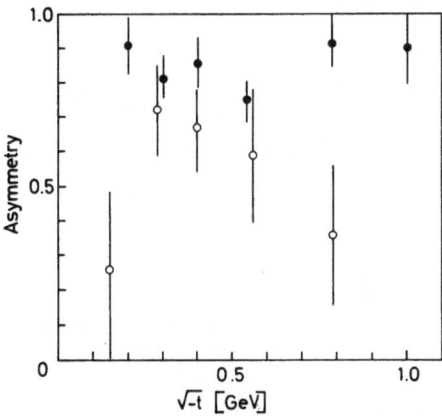

Fig. 3. Comparison of $|A_{\pi^+}|$ (open circles) at 16 GeV/c (Ref. [44]), to $\frac{1}{2}(1 + \Sigma_{\pi^+})$ (closed circles) at 12 GeV/c (Ref. [42])

It is interesting to note that for $t \approx -0.1$ GeV^2 the bound is almost saturated which can only come about if the M_{10}^+ and M_{11}^+ amplitudes are orthogonal to each other.

Isospin for πN Photoproduction

The isospin decomposition which is most suitable when one considers t channel exchanges is given in Ref. [45] and quoted here for completeness

$$
\begin{aligned}
M(\gamma p \to \pi^+ n) &= \sqrt{2} M^{(-)} + \sqrt{2} M^{(0)} \\
M(\gamma n \to \pi^- n) &= -\sqrt{2} M^{(-)} + \sqrt{2} M^{(0)} \\
M(\gamma p \to \pi^0 p) &= \qquad\qquad M^{(0)} + M^{(+)} \\
M(\gamma n \to \pi^0 n) &= \qquad\qquad -M^{(0)} + M^{(+)} .
\end{aligned}
\tag{40}
$$

Construction of Models

Photoproduction of π^0

By now we have established the general formalism and may start looking at models. Probably the simplest case is π^0 photoproduction, so let us try to build a model from scratch and as far as possible let the experiments tell us what ingredients we need.

As is seen from Table 2, the well-established particles that may be exchanged are all spin one particles. The simple Born term with elementary particles exchanged will, therefore, give a differential cross-section

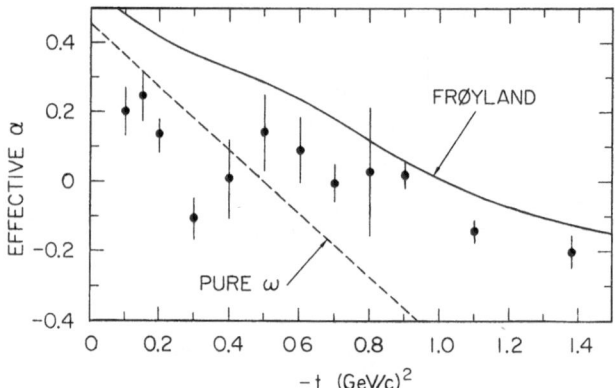

Fig. 4. α_{eff} for $\gamma p \to \pi^0 p$. From Ref. [39]

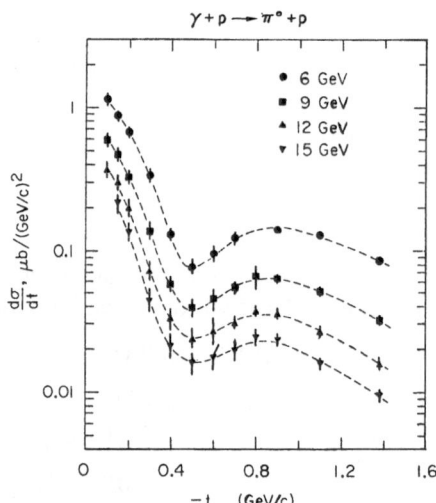

Fig. 5. $d\sigma/dt$ for $\gamma p \to \pi^0 p$. From Ref. [39]

independent of energy. Experimentally, there is a fairly rapid decrease with energy [39], (Fig. 4) and hence the necessity to Reggeize.

The general shape of the differential cross-section [39], Fig. 5, gives a strong hint that the process is dominated by ω exchange. The ω exchange is supposed to contribute dominantly to amplitudes that do not change the nucleon helicity, i.e., $n = 1$ amplitudes. From pure kinematics it then follows that there will be a forward dip, and, furthermore, this dip will remain after absorption. Also the dip at $t = -0.5$ GeV^2 agrees nicely with the Regge picture which predicts a dip at the nonsense wrong signature point $\alpha_\omega(t) = 0$ due to the vanishing of the signature factor at this point. In the Michigan model [2] the dip comes about as the result of interference between the pole and the cut term, but the arguments leading to ω exchange dominance are independent of a particular dip mechanism.

Our model will predict a nice clean dip in the forward direction. Experimentally [46], however, the differential cross-section rises very steeply as one approaches the extreme forward direction. This is due to the exchange of a γ; the Primakoff effect. Because of the factor $1/t$ in the photon propagator this term becomes dominant at sufficiently small $|t|$ in spite of the smallness of the electromagnetic coupling of the γ to the proton. Notice that $t = 0$ is just outside the physical region. Given the π^0 lifetime there are no unknowns in the calculation of the Primakoff effect. Since the virtual γ couples to the electric charge essentially in the nucleon helicity non-flip amplitudes it follows that $n = 1$ for the amplitudes where the γ exchange is important. This again means that also the Primakoff effect has to vanish in the exact forward direction. The photon has spin one and mass zero. Therefore, the one-photon exchange has $\alpha = 1$ in Regge language. Nevertheless, the Primakoff spike will continue to grow with energy, a fact that can easily be checked from the parametrization, in Eq. (41), and the fact that $t_0 \approx -m^2 \mu^4/s^2$. Because the width of the Primakoff spike is proportional to t_0 it is seen to shrink very rapidly, with energy. However, the integral under the peak grows like $\ln s$. In π^0 photoproduction this will probably be impossible to detect because of the narrowness and rapid shrinkage of the spike. However, in photoproduction of η the Primakoff peak should be detectable up to much higher energies because of the big mass of the η.

Though it may be difficult to observe the Primakoff peak directly at high energies one might make some interesting observations of the γ exchange through its interference with the hadronic part of the exchanges which is dominant a bit away from the forward direction. This interference term is proportional to

$$t^{-1}(-t + t_0)s \, \mathrm{Re}\,\omega \approx -s \, \mathrm{Re}\,\omega$$

because of the smallness of t_0. Therefore, the interference term will behave more or less as a simple decreasing (with t) background, but which is more slowly falling with energy than the purely hadronic exchange contributions. Already at $5-6$ GeV/c the interference term may constitute about 15% of the differential cross-section near the maximum at $t = -0.1$. From this it is quite clear that the Primakoff effect should not be left out in realistic calculations. Clearly, if the Primakoff effect is included in the analysis it makes the effective α of the hadronic exchanges smaller if the interference is constructive and bigger if the interference is destructive.

A recent and accurate determination [39] of the total effective α shows that it is embarrassingly low (Fig. 4), a possible indication of the interference being destructive. On the other hand, the assumption of destructive interference may create difficulties for the $-t$ dependence, but to the author's knowledge no analysis has yet been carried out with such assumptions.

Since our model so far only has contributions from natural parity exchanges and only to the $n = 1$ amplitudes we immediately predict that $\Sigma = 1$ even if absorption is used. Since Σ_{π^0} has been measured [39, 40] (Fig. 2) we know that this is not true, so we need something in the two other amplitudes. The candidates are either ϱ or B or both. This type of contribution must also be there because of the known difference between π^+ and π^- photoproduction off nucleons. The B would make $\Sigma \neq 1$ simply by being an unnatural parity object and the ϱ exchange would make $\Sigma \neq 1$ through its ϱP cut which has both natural and unnatural parity parts. If the B trajectory has anything close to the universal slope one should be able to tell which mechanism is in operation through the energy dependence. If B exchange dominated there would be a fairly rapid approach to $\Sigma = 1$, while if ϱP is dominant one would expect a very slow energy dependence. Another point is that if B exchange is dominant then the polarization will be roughly proportional to the BP cut which again should show up in the energy dependence of the polarized target asymmetry.

Obviously our model by now will predict a ratio

$$R_{\pi^0} \equiv d\sigma(\gamma n \rightarrow \pi^0 n)/d\sigma(\gamma p \rightarrow \pi^0 p) = 1$$

apart from the very forward direction where the absence of the Primakoff peak in $\gamma n \rightarrow \pi^0 n$, makes the ratio smaller. R_{π^0} has been measured [47] and found to be less than one. This is not surprizing since both the magnetic and the electric coupling of the ω and ϱ mesons to the $\bar{N}N$ are not zero at the poles. Therefore, we have to introduce a small contribution of the ω to the $n = 0$ and $n = 2$ amplitudes, and also a contribution of the ϱ to the $n = 1$ amplitudes.

We therefore arrive at the following *Regge pole parametrizøn* for photoproduction of neutral pions from protons.

$$
\begin{aligned}
M^R_{0-\frac{1}{2};1\frac{1}{2}} &= t(\gamma^\omega_{11}\omega + \gamma^\varrho_{11}\varrho + \gamma^B_{11}B) \\
M^R_{0\frac{1}{2};1-\frac{1}{2}} &= t(-\gamma^\omega_{11}\omega - \gamma^\varrho_{11}\varrho + \gamma^B_{11}B) \\
M^R_{0\frac{1}{2};1\frac{1}{2}} &= \sqrt{-t+t_0}(\gamma^\omega_{10}\omega + \gamma^\varrho_{10}\varrho + (-t)^{-1}e\gamma_{\pi^0}s) \\
M^R_{0-\frac{1}{2};1-\frac{1}{2}} &= M_{0\frac{1}{2};1\frac{1}{2}}.
\end{aligned}
\tag{41}
$$

Here

$$t_0 \approx -m^2 \mu^4/s^2$$
$$\omega = \Gamma(1 - \alpha_\omega)(s/s_0)^{\alpha_\omega} \xi(\alpha_\omega)$$
$$\varrho = \omega(\alpha_\omega \to \alpha_\varrho)$$
$$B = \Gamma(-\alpha_B)(s/s_0)^{\alpha_B} \xi(\alpha_B)$$
$$\xi(\alpha) = \begin{cases} (1 - e^{-i\pi\alpha}) & \text{(Argonne model)} \\ i\,e^{-i\pi\alpha/2} & \text{(Michigan model)} \end{cases}$$

and γ_{π^0} is simply related to the π^0 lifetime. The rest of the γ's are residue functions containing possible α and t dependent factors. The factor $\Gamma(-\alpha_B)$ occurs when the B is taken to be exchange degenerate with the π in which case the B does not have to choose nonsense at $\alpha_B = 0$.

The signature factor in the Michigan model does not vanish at $\alpha_B = 0$, so to avoid a pole there there has to be a change of the factor $\Gamma(-\alpha_B)$. For instance, $\Gamma(-\alpha_B) \to \Gamma(\frac{1}{2}(1 - \alpha_B))$. The final step to put the model to work is then to perform the absorption according to some prescription, for instance Eq. (20), or to introduce cuts in some more phenomenological way.

The presence of the ϱ (or B), in the $n = 0$ and $n = 2$ amplitudes is seen dramatically in the photoproduction of charged pions where ϱ quantum number (or again possibly B quantum number) exchange is responsible for the smallness of the π^-/π^+ ratio.

As no particular structure is seen in this ratio around $t \approx -0.5\ \mathrm{GeV}^2$ one may conclude that the effective ϱ (or B) contribution has no structure in the $n = 0$ plus $n = 2$ amplitudes. Since the ϱP cut from these amplitudes has an unnatural parity part, one expects $d\sigma^{\parallel}/dt$ in π^0 photoproduction to be structureless around $t \approx -0.5\ \mathrm{GeV}^2$. This is indeed the case as seen from Fig. 6. From what we have learned the dip structure in $\pi^0 p$ photoproduction arises from the $n = 1$ amplitude which has only natural parity exchange. Since the data [47] for R_{π^0} (Fig. 7) indicate a weak minimum around $t \approx -0.5\ \mathrm{GeV}^2$ this is most easily interpreted as an enhancement of $n = 1$ amplitudes relative to $n = 0$ and $n = 2$ amplitudes in going from $\pi^0 p$ to $\pi^0 n$ photoproduction. This immediately tells us that

$$\Sigma_{\pi^0 n} > \Sigma_{\pi^0 p}. \tag{42}$$

By similar arguments one finds

$$|A_{\pi^0 n}| < |A_{\pi^0 p}|. \tag{43}$$

Again these conclusions are not necessarily correct for very small t because of the absence of the Primakoff effect in the $\pi^0 n$ case so that $\Sigma_{\pi^0 n}$ will go nicely to zero in the forward direction while $\Sigma_{\pi^0 p} \to 1$ for the smallest angles attainable because of finite resolutions. Preliminary data [48] for $A_{\pi^0 p}$ show that it is quite large and negative, in general agreement with most models.

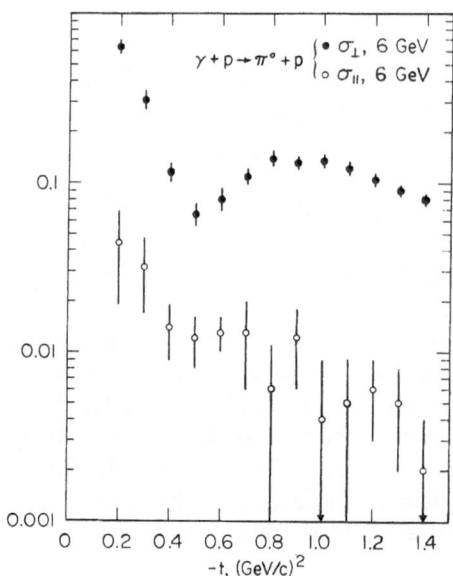

Fig. 6. $d\sigma^\perp/dt$ (closed circles) and $d\sigma^{\parallel}/dt$ (open circles) at 6 GeV/c. From Ref. [39]

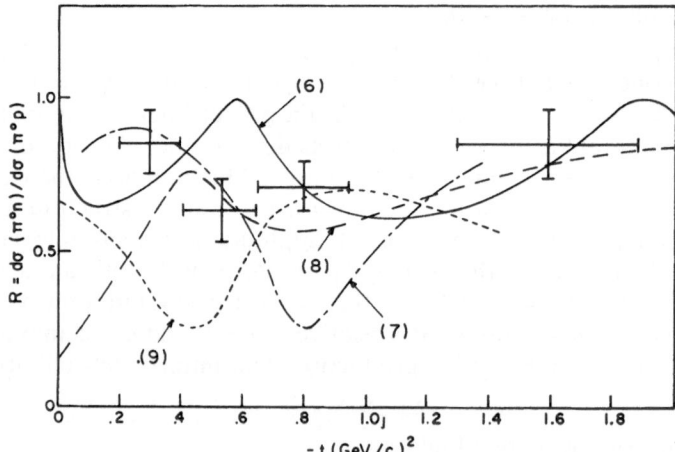

Fig. 7. Ratio $(\gamma n \to \pi^0 n)/(\gamma p \to \pi^0 p)$. From Ref. [47]

η Photoproduction

Because η is an isoscalar and π^0 and isovector, the role of the isoscalar and isovector parts of the current are reversed in the cases of η and π^0 photoproduction.

Using SU_3 and the vector dominance model we have for the upper vertex [49]

$$g_{\gamma\eta\omega}/g_{\gamma\eta\varrho} = \tfrac{1}{3} \tag{44}$$

while for π^0 photoproduction

$$g_{\gamma\pi\omega}/g_{\gamma\pi\varrho} = 3 \tag{45}$$

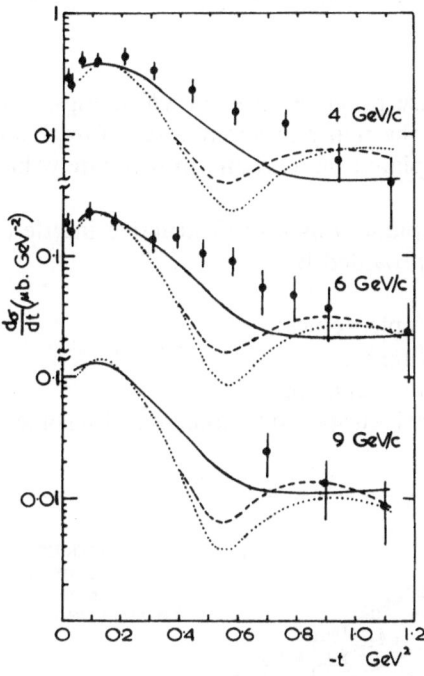

Fig. 8. $d\sigma/dt$ for $\gamma p \to \eta p$. From Ref. [3] Data from Ref. [50]

so that in η photoproduction the importance of ϱ exchange is enhanced a factor 9 relative to π^0 photoproduction. On the general, semi-experimental grounds given previously, we then expect the η photoproduction to have little or no structure around $t = -0.5\ \mathrm{GeV}^2$. This is indeed the case experimentally [50, 51]. In actual models this lack of structure comes about in a natural way in the Michigan model, while the NWS zero from weak cut models can be filled by the B meson exchange, but the energy dependence of the two types of models is quite different because the B trajectory is low lying. The DESY and SLAC data [50] (Fig. 8) are in best agreement with no B exchange, while the Cornell data [51] agree with the B exchange model. A detailed discussion of these points can be found in Refs. [3] and [4].

Maybe it is not surprising that the ϱ behaves like an irregular object in view of the very big difficulties that simple models have had in fitting πN CEX data. In fact, in a recent paper [1] the amplitudes have been reconstructed from existing data, and the only model consistent with these amplitudes is a $\varrho + \varrho'$ model [52].

Because the ϱ contribution introduces a lot of unnatural parity via the absorption, we expect Σ_η to be substantially smaller than Σ_{π^0}.

$\pi^\pm$ Photoproduction

Typical data for these processes are shown in Figs. 9–13. The analysis is made difficult by the many different ways of explaining the forward spike [23]. The explanation chosen unfortunately has consequences also at larger angles.

The principal explanations are: destructive interference of the pole with a background provided by

1. cuts;
2. conspiring complex poles;
3. direct channel terms;
4. unknown background; or
5. the π trajectory is supposed to conspire with some other trajectory.

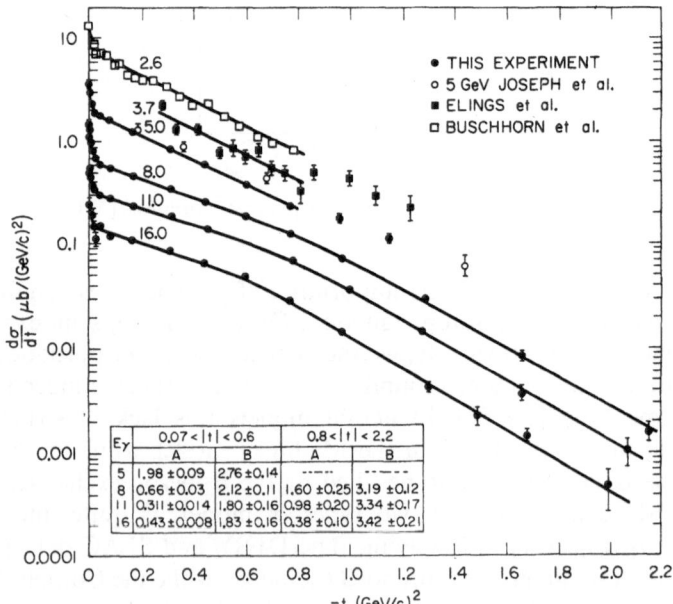

Fig. 9. $d\sigma/dt$ for $\gamma p \to \pi^+ n$. From Ref. [23]

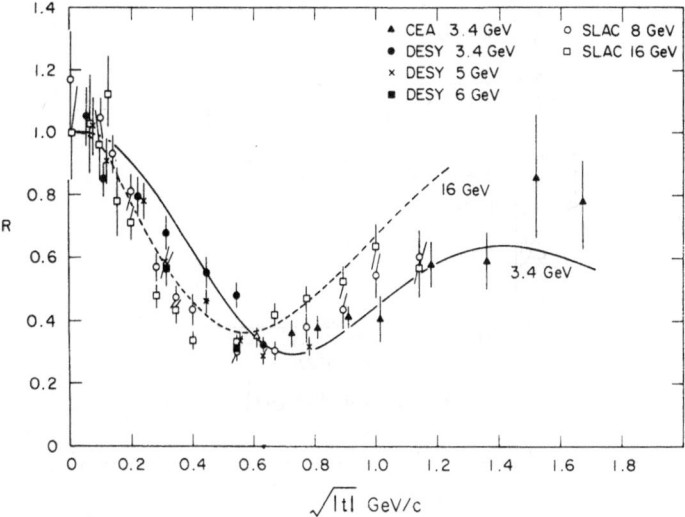

Fig. 10. Ratio $(\gamma n \to \pi^- p)/(\gamma p \to \pi^+ n)$. From Ref. [23]

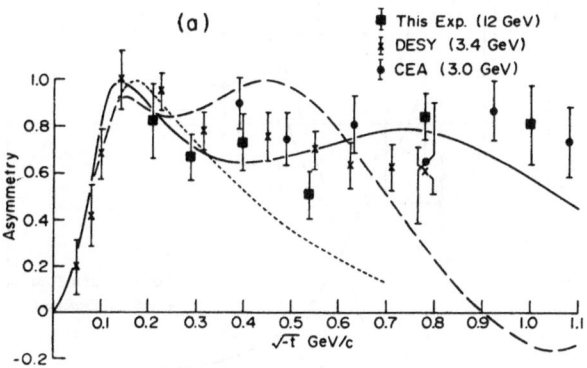

Fig. 11. Σ_{π^+} from Ref. [42]

The last explanation can probably be ruled out because factorization then would demand [53] that there is a forward dip in

$$\pi^+ p \to \varrho^0 \Delta^{++}$$

whereas a peak has been observed [54]. Using the explanation of the direct channel effects as the background, one immediately admits defeat in $np \to pn$ which has a spike [55] similar to the one seen in charged pion photoproduction. In that case there is no direct channel resonances to

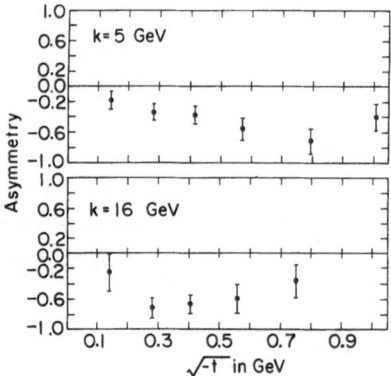

Fig. 12. A_{π^+} from Ref. [44]

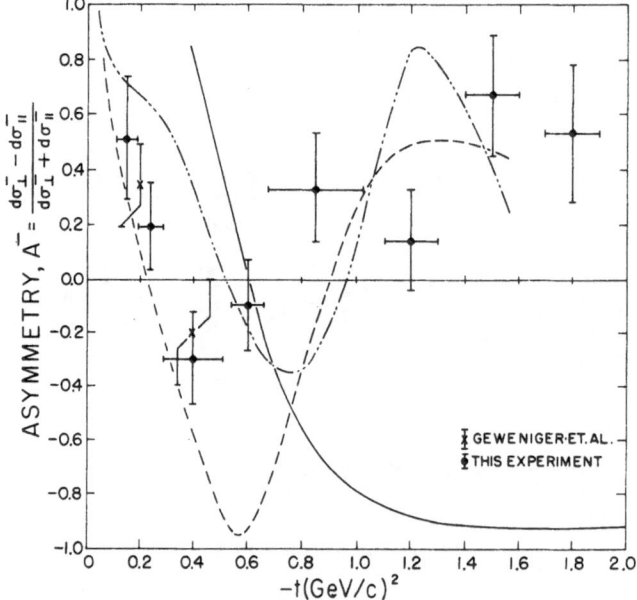

Fig. 13. Σ_{π^-} from Ref. [65]

help[3]. Anyhow, it is necessary to multiply the Born term with a quite strongly t dependent form factor [56] in order that the cross-section shall not become completely unreasonable at $-t > \mu^2$. The meaningfulness of such a procedure may, however, be questionable.

---

[3] Of course, there is the π^0 in the u channel which might play the same role as the nucleon in photoproduction, but many good arguments can be given against such an explanation.

The complex poles clearly can fit the data, but the physical interpretation of such fits is difficult, so we shall concentrate on the cut explanation.

Let us start by considering only π exchange and call the pion contribution to the $n = 0$ amplitude

$$M_\pi(t) \sim t/(t - \mu^2) \tag{46}$$

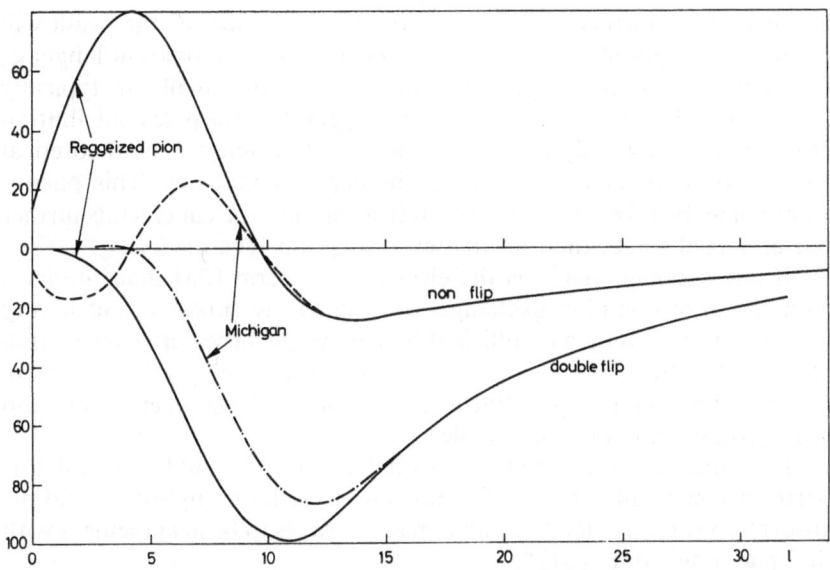

Fig. 14. Partial wave amplitudes times $(2J + 1)$ for the real part of the unabsorbed Reggeized π non-flip and double-flip amplitudes compared to the same for the Michigan model for the process $np \to pn$. From Ref. [57]

so that $M_\pi(0) = 0$. Approximating $d^J_{\frac{1}{2}\frac{1}{2}}(\theta)$ by $P^l(\cos\theta)$ we obviously must have

$$\sum_l (2l + 1)\, a^l_\pi(s) = 0\,. \tag{47}$$

The highest partial waves will be negative and depend very little on what form factors one uses for the π exchange, while the lowest partial waves will be positive and strongly dependent on form factors, but in such a way that

$$\sum_{\text{positive } a_l} (2l + 1)\, a^l_\pi$$

does not depend much on the form factor. A figure displaying $(2l + 1)a^l$ as a function of l for Reggeized one-pion exchange in np CEX is shown in Fig. 14. What the absorption model does is to subtract something from the lowest partial waves and as a result one gets a non-vanishing cross-section at $t = 0$. However, if $\lambda_\pi = 1$, there is still too much left of the positive part of the partial waves, i.e., the cross-section is too small. Therefore, the Michigan model increases λ_π until the forward differential cross-section is correct. As is seen from Fig. 14, the lowest partial waves are now negative and not very small and the sum of the lowest 10 or so partial waves is roughly zero. Clearly, any model that wants to fit the data must have a sum of the lowest partial waves that is small. But, if this is achieved by a cancellation like in the Michigan model, the result will be that the amplitude becomes big again at larger t. In pole cut language it means that the pole will cancel the cut at some small $-t$, typically $-t \approx \mu^2$ but the cut contribution will be bigger than the pole contribution at larger $|t|$, and, finally become so big that it dominates the theoretical cross-section which is bigger than the experimental one. This phenomenon may be taken as a possible indication that the cut extends further out in partial waves than in present absorption type models.

If one assumes that it is the electric Born term [23] that interferes with the pure one-pion exchange, one effectively subtracts out a very large s wave which then is sufficient to achieve the necessary cancellation of the rest of the low partial waves. However, the low partial waves are now very big and if no form factor is used one will again get a much too large cross-section at larger angles.

It seems, therefore, that a reasonable model should have all low partial waves small not to be in conflict with the large angle data, and the absorption should extend to higher partial waves to be in agreement with the small angle data [57][4].

Clearly, the question of how the πP cut behaves at larger t is very important because an incorrect πP cut will distort the other contributions, especially the A_2 contributions to the $n = 0$ and $n = 2$ amplitudes that can hardly be distinguished from the πP cut except through their energy dependence and corresponding phases. Therefore, one should not consider present day fits as anything final.

All models that want to fit the data have to have a very small $n = 0$ amplitude at $|t| > \mu^2$. At larger angles the double flip amplitude must be dominant at least in π^+ photoproduction because of the large π exchange contribution to this amplitude which is not much changed by absorption. Around $t = -0.25$ the ratio $R = d\sigma(\gamma n \to \pi^- p)/d\sigma(\gamma p \to \pi^+ n)$ is small and, therefore, there has to be a strong interference between the $\varrho + B$ types of exchanges and the $\pi + A_2$ types of exchanges in the $n = 2$ amplitude. Since the absorption is unimportant in this amplitude one may think in terms of pure Regge poles. Taking ϱ, A_2 to be strongly exchange

---

[4] After this talk was given we have received a preprint [11] by Worden who has successfully used an approach like the one suggested here.

degenerate the phase of the $\varrho + A_2$ contribution may be $e^{-i\pi\alpha}$ in the π^+ case and 1 in the π^- case [5]. This makes possible the strongly destructive interference with the π exchange contribution that we need to explain R. At the nonsense wrong signature point of the ϱ obviously the $n = 2$ amplitudes for π^+ and π^- production are roughly the same. This means that also the $n = 0$ plays an important role in explaining R but the interferences between the various poles and their cuts are now quite complicated. The cuts in the $n = 0$ amplitude are proportional to the corresponding pole strengths, and since the natural parity contributions have opposite signs in the $n = 0$ and $n = 2$ amplitudes, while the unnatural parity contributions have the same sign, we predict $R > 1$ for $t = 0$ and $R = 1$ at about $t = -\mu^2$. Experimentally, this seems to be well satisfied (Fig. 10).

In the Michigan model the ϱ contribution does not vanish at $\alpha_\varrho = 0$ and, therefore, the link between the ϱ and A_2 contribution becomes weaker than in a strong exchange degeneracy model. Also in the Michigan model the details of the various interferences are quite complicated. The interference between the π exchange and the πP cut almost extinguishes the unnatural parity contributions close to $t = -\mu^2$. Accordingly, we expect $\Sigma_{\pi^+} \approx 1$ at this point which is indeed observed [43].

The asymmetry with a polarized target has been measured at 5 GeV/c and 16 GeV/c and is large and negative [44]. Obviously, therefore, the $n = 1$ amplitudes are non-zero. Candidates to contribute are the $\varrho + A_2$ which we expect to contribute since both πN CEX and $\pi N \to \eta N$ do not vanish in the forward direction.

It is easy to see from Eqs. (26) and (35) that the polarization gets its main contributions from the interference between πP and the $\varrho + A_2$ contributions to the $n = 1$ amplitudes. At small angles these amplitudes will be almost orthogonal to each other and this permits the polarization to become quite big. However, the phase of the $(\varrho + A_2)$ contribution will rotate faster with t (how much faster depends on details of the models) than the πP cut and therefore one expects the polarization to become small again. The most naïve guess is that it would be zero at an angle somewhat larger than that which corresponds to the NWS zero of the ϱ contribution. Experimentally it is still sizeable at $\alpha_\varrho(t) = 0$.

The dip mechanism in the Michigan model and also Harari's model [58] may be said to be kinematical or geometrical since the absence or presence of a dip is supposed to depend only on the spin structure of the amplitudes. The Argonne type of models may be said to have dynamical dips. The presence of dips depending only on the particles exchanged and not on the spin states. Harari [59] has recently proposed a test of the two

---

[5] A slightly different result is obtained assuming the absence of exotics in $K^*N \to KN$ and using SU_3 and the vector dominance model.

types of dip mechanism by comparing π^0 photoproduction to π^0 electro-production. If the dip mechanism is of geometrical nature, the dip should go away with increasing $|q^2|$ for the virtual photon.

Since good phase shift analysis exists for low energy photoproduction, one may use finite energy sum rules (FESR) or continuous moment sum rules (CMSR) to obtain predictions for high energies [13, 60]. This approach is a check on the duality idea, and secondly it may simplify the fitting procedure only to fit the FESR and then predict the high energy results, which, if in agreement with data, would give us confidence in the high energy model we were testing.

Such a test has been carried out for the very forward direction for π^+ photoproduction [13]. The result was that, at present, the FESR could not distinguish between a conspiracy model and an absorptive cut model. However, one may expect the FESR to become a more sensitive test as more accurate phase shifts become available [6].

Photoproduction of K mesons

The processes best studied are

$$\gamma p \to \begin{cases} K^+ \Lambda \\ K^+ \Sigma^0 . \end{cases}$$

The particles that may be exchanged are K, K^* and K^{**} analogous to π, ϱ and A^2 in charged pion photoproduction. Naïvely, one might then expect a shape similar to that which is seen in charged photoproduction with the proper corrections for the π and K mass differences. Experiments [61] (Fig. 15), however, show a quite different behaviour with a dip in the forward direction, a maximum at about $t = -0.2 \, \text{GeV}^2$ and then a smooth fall off with increasing angle.

A very simple but nice analysis has been performed in Ref. [62]. K exchange is neglected, and since a dip is observed this indicates $K^* - K^{**}$ dominance in the $n = 1$ amplitudes. From the Λ/Σ^0 ratio at $|t| > 0.2 \, \text{GeV}^2$ F/D ratios for the baryon non-flip amplitudes are obtained in good agreement with other data. The Λ/Σ^0 ratio at $t = 0$ is determined by the $n = 0$ amplitude and is supposedly dominated by $P(K^* + K^{**})$ cuts. Since the strength of the cuts are proportional to the strength of the poles one can also deduce the F/D ratio for the baryon flip amplitude; correct to the extent one can neglect K exchange. The F/D ratio thus obtained is in agreement with other data.

Measurements with a polarized target [14] has been done at SLAC without distinguishing between the Λ and the Σ^0. The result is a quite large and negative polarization. This looks a bit embarrassing to the

---

[6] Several new FESR tests have been carried out in Ref. [11].

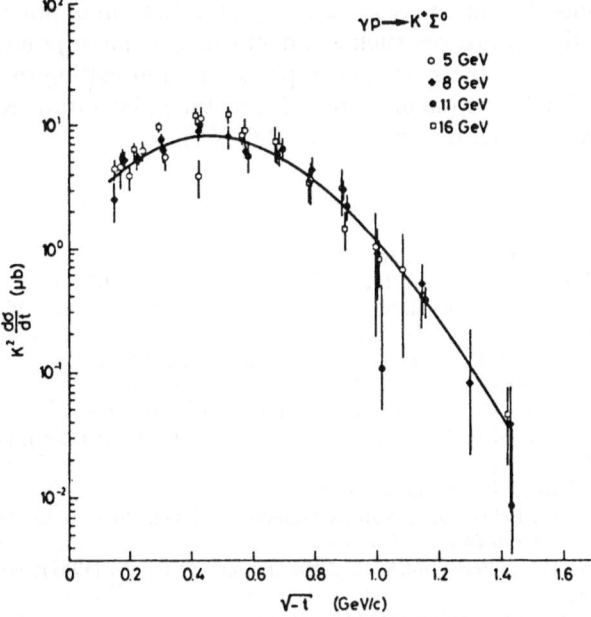

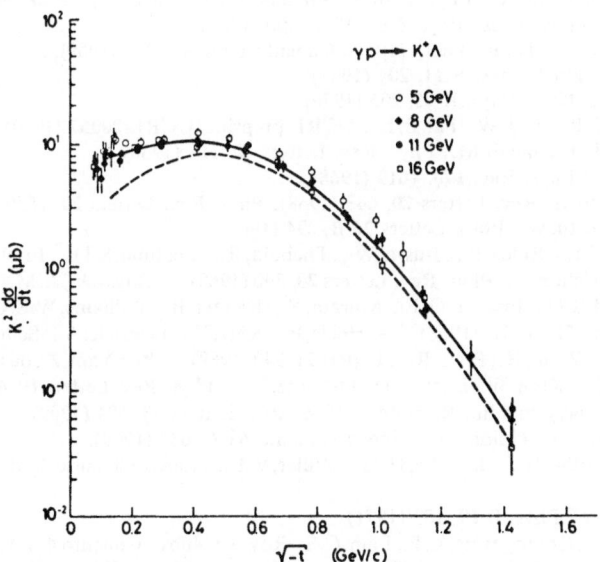

Fig. 15. $k^2 d\sigma/dt$ versus $\sqrt{-t}$ for $\gamma p \to K^+ \Sigma^0$ and $p \to K^+ \Lambda$. From Ref. [62]. Data from Ref. [61]

$K^* + K^{**}$ model that one may expect to give only a small polarization. However, in other processes such a model can give large polarizations [63]. It might be that the K exchange plays a role in getting the polarization correct, but no actual attempt at fitting the polarization is known. For a more detailed discussion, see Ref. [64].

References

1. Halzen, F., Michael, C.: Phys. Letters **36** B, 367 (1971).
2. Henyey, F., Kane, G. L., Pumlin, F., Ross, M. H.: Phys. Rev. **182**, 1579 (1969). — Kane, G. L., Henyey, F., Richards, P. R., Ross, M., Williamson, G.: Phys. Rev. Letters **25**, 1519 (1970).
3. Gault, F. D., Martin, A. D., Kane, G. L.: Nucl. Phys. B **32**, 429 (1971).
4. Goldstein, G. R., Owens, J. F. III: Tufts University preprint (1971).
5. Kramer, G., Schilling, K., Stodolsky, L.: Nucl. Phys. B **5**, 317 (1968).
6. Blackmon, M. L., Kramer, G., Schilling, K.: Nucl. Phys. B **12**, 495 (1969); Phys. Rev. **183**, 1452 (1969).
7. Colocci, M.: Nuovo Cimento Letters **4**, 53 (1970).
8. Contogouris, A. P., Lebrun, J. P., von Bochmann, G.: Nucl. Phys. B **13**, 246 (1969).
9. — — Nuovo Cimento **64** A, 627 (1969).
10. Dar, A., Watts, T. L., Weisskopf, V. F.: Phys. Letters **30** B, 264 (1969); Nucl. Phys. B **13**, 477 (1969).
11. Worden, R.: CALTECH preprint CALT-68-313.
12. Benfatto, G., Nicolo, F., Rossi, G. C.: Nuovo Cimento Letters **1**, 537 (1969); Nuovo Cimento **64** A, 1033 (1969).
13. Jackson, J. D., Quigg, C.: Phys. Letters **29** B, 236 (1969); Nucl. Phys. B **22**, 307 (1970).
14. Frøyland, J., Gordon, D.: Phys. Rev. **177**, 2500 (1969).
15. Capella, A., Tran Thanh Van, J.: Nuovo Cimento Letters **1**, 321 (1969); **3**, 1199 (1969).
16. Frøyland, J.: Nucl. Phys. B **11**, 204 (1969).
17. Kellett, B. H.: Nucl. Phys. B **25**, 205 (1970).
18. Ball, J. S., Muller, H. J. W., Pal, B. K.: UCRL preprint, UCRL-20057 (1970).
19. — Frazer, W. R., Jacob, M.: Phys. Rev. Letters **20**, 518 (1968).
20. Henyey, F. S.: Phys. Rev. **170**, 1619 (1968).
21. Cooper, F.: Phys. Rev. Letters **20**, 643 (1968); Phys. Rev. Letters **20**, 1550 E (1968).
22. Dietz, K., Korth, W.: Phys. Letters **26** B, 394 (1968).
23. Boyarski, A. M., Bulos, F., Busza, W., Diebold, R., Ecklund, S. D., Fischer, G. E., Rees, J. R., Richter, B.: Phys. Rev. Letters **20**, 300 (1968). — Boyarski, A. M., Diebold, R., Ecklund, S. O., Fischer, G. E., Murata, Y., Richter, B., Williams, W. S. C.: Phys. Rev. Letters **21**, 1767 (1968). — Heide, P., Kätz, V., Lewis, R. A., Schmüser, P., Skronn, H. J., Wahl, H.: Phys. Rev. Letters **21**, 248 (1968). — Bar-Yam, Z., de Pagter, J., Hoenig, M. M., Kern, W., Luckey, D., Osborne, L. S.: Phys. Rev. Letters **19**, 40 (1967).
24. Ahmad, M., Fayyazuddin, Riazuddin: Phys. Rev. Letters **23**, 504 (1969).
25. Ademollo, M., Del Guidice, E.: Nuovo Cimento **63** A, 639 (1969).
26. Bender, I., Rothe, H. J., Dosche, H. G., Müller, V. F.: Nuovo Cimento Letters **3**, 625 (1970).
27. Reya, E.: Nucl. Phys. B **29**, 189 (1971).
28. Argyres, E. N., Contogouris, A. P., Lam, C. S., Roy, S.: Nuovo Cimento **4** A, 156 (1971).
29. Bender, I., Dosch, H. G., Rothe, H. J.: Nuovo Cimento **62** A, 1025 (1969).
30. Boyarski, A. M., Diebold, R., Ecklund, S. D., Fischer, G. E., Murata, Y., Richter, B., Sands, M.: Phys. Rev. Letters **25**, 695 (1970), erratum p. 1148.
31. — — — — Murata, Y., Richter, B., Sands, M.: Phys. Letters **34** B, 547 (1971).

32. Halzen, F., Mandula, J., Weyers, J., Zweig, G.: CERN preprint TH. 1371 (1971).
33. Baglin, C., Briandet, P., Carlson, P. J., Chabaud, V., Davier, M., Eide, A., Fleury, P., Gracco, V., Johansson, E., Lehmann, P., Lundby, A., Morand, R., Mukhin, S., Myrheim, J., Treille, D.: Paper presented at the Amsterdam conference (1971).
34. Michael, C.: Phys. Letters 29 B, 230 (1969).
35. Sopkovich, N. J.: Nuovo Cimento 26, 186 (1962). — Gottfried, K., Jackson, J. D.: Nuovo Cimento 34, 735 (1964).
36. Gell-Mann, M., Goldberger, M. L., Low, F. E., Marx, E., Zachariasen, F.: Phys. Rev. 133, B 145 (1964).
37. Cohen-Tannoudji, G., Morel, A., Navelet, H.: Ann. Phys. (N.Y.) 46, 239 (1968).
38. Stichel, P.: Z. Physik 180, 170 (1964). — Ader, J. P., Capdeville, M., Cohen-Tannoudji, G., Salin, Ph.: Nuovo Cimento 56 A, 952 (1968).
39. Anderson, R. L., Gustavson, D. B., Johnson, J. R., Overman, I. D., Ritson, D. M., Wiik, H. B.: Phys. Rev. Letters 26, 30 (1971). — Anderson, R. L., Gustavson, D. B., Johanson, J. R., Overman, I. D., Ritson, D. M., Wiik, B. H., Worcester, D.: SLAC preprint, SLAC-PUB-925 (1971).
40. Bellenger, D., Bordelon, R., Cohen, K., Deutsch, S. B., Lobar, W., Luckey, D., Osborne, L. S., Pothiev, E., Schwitters, R.: Phys. Rev. Letters 23, 540 (1969).
41. Geweniger, C., Heide, P., Kötz, U., Lewis, R. A., Schmüser, P., Skronn, H. J., Wahl, H., Wegener, K.: Phys. Letters 29 B, 41 (1969); 28 B, 155 (1969).
42. Schwitters, R. F., Leony, F., Luckey, D., Osborne, L. S., Boyarski, A. M., Ecklund, S. D., Siemann, R., Richter, B.: Phys. Rev. Letters 27, 120 (1971).
43. Burfeindt, H., Buschhorn, G., Geweniger, C., Heide, P., Kotthaus, R., Wahl, H., Wegener, K.: Phys. Letters 33 B, 509 (1970).
44. Morehouse, C. C., Borghini, M., Chamberlain, O., Fuzey, R., Gorn, W., Powell, T., Robrish, P., Rock, S., Shannon, S., Shapiro, G., Weisberger, H., Boyarski, A., Ecklund, S., Murata, Y., Richter, B., Siemann, R., Diebold, R.: Phys. Rev. Letters 25, 835 (1970).
45. Chew, G. F., Goldberger, M. L., Low, F. E., Nambu, Y.: Phys. Rev. 106, 1345 (1957).
46. Braunschweig, M., Braunschweig, W., Husman, D., Lübelsmeyer, K., Schmitz, D.: Phys. Letters 26 B, 405 (1968); Nucl. Phys. B 20, 191 (1970).
47. Bolon, G. C., Bellenger, D., Lobar, W., Luckey, D., Osborne, L. S., Schwitters, R.: Phys. Rev. Letters 27, 964 (1971).
48. Booth, P. S. L.: University of Liverpool, presented at the Cornell conference (1971).
49. Dar, A., Weisskopf, V. F.: Phys. Letters 26 B, 405 (1968).
50. Johnson, J. R.: SLAC Report No. 124 (1970). — Braunschweig, W., Braunschweig, M., Frese, H., Lübelsmeyer, K., Meyer-Wachsmuth, H., Schmitz, D., Schultz von Dratzig, A., Wessels, G.: Phys. Letters 33 B, 236 (1970).
51. Soh, E.: Cornell. Private communications to Ref. [23].
52. Barger, V., Phillips, R. J. N.: Phys. Rev. 187, 2210 (1969).
53. Le Bellac, M.: Phys. Letters 25 B, 524 (1967).
54. Aderholz, M., Bartch, J., Deutschmann, M., Kraus, G., Speth, G., Grote, C., Lanius, K., Nowak, S., Walter, M., Böttcher, H., Byer, T., Cocconi, V. T., Hansen, J. P., Kellner, G., Mihul, A., Morrison, D. R. O., Moskalev, V. I., Töfte, H.: Phys. Letters 27 B, 174 (1968).
55. Engler, J., Horn, K., Monning, F., Schuludecker, P., Schmidt-Parzefall, W., Schopper, H., Sievers, P., Ullrich, H., Hartung, R., Runge, K., Galaktinov, Yu.: Phys. Letters 34 B, 528 (1971).
56. Kellett, B. H.: Nucl. Phys. B 25, 205 (1970).
57. Frøyland, J., Winbow, G. A.: To be published in Nucl. Phys.
58. Harari, H.: Phys. Rev. Letters 26, 1400 (1971).
59. — Phys. Rev. Letters 27, 1028 (1971).

60. Bietti, A., Di Vecchia, P., Drago, F., Paciello, M. L.: Phys. Letters **26** B, 457 (1968). — Di Vecchia, P., Drago, F., Ferro-Fontan, C., Odorico, R., Paciello, M. L.: Phys. Letters **27** B, 296 (1968). — Di Vecchia, P., Drago, F., Ferro-Fontan, C., Odorico, R.: Phys. Letters **27** B, 521 (1968). — Di Vecchia, P., Drago, F., Paciello, M. L.: Nuovo Cimento **55** A, 809 (1968). — Chu, S. Y., Roy, D.: Phys. Rev. Letters **20**, 958 (1968); Phys. Rev. **171**, 1762 (1968).
61. Boyarski, A., Bulos, F., Busza, W., Diebold, R., Ecklund, S., Fischer, G., Rees, J., Richter, B.: Phys. Rev. Letters **22**, 1131 (1969).
62. Michael, C., Odorico, R.: Phys. Letters **34** B, 422 (1971).
63. Irving, A. C., Martin, A. D., Michael, C.: Nucl. Phys. B **32**, 1 (1971).
64. Capella, A., Tran Thanh Van, J.: Nuovo Cimento Letters **41**, 1199 (1970).
65. Bar-Yam, Z., de Pagter, J., Dowd, J., Kern, W.: Phys. Rev. Letters **24**, 1078 (1970).

Dr. J. Frøyland
Institute of Physics
Oslo 3, Norway

Some Aspects of Vector Meson Photoproduction on Protons

K. Schilling

Contents

Abstract

Some recent models on diffraction dissociation of photons are reviewed that go beyond simple vector meson dominance of the final state by including non resonant diffractively produced background.

1. Introduction

The high energy photoproduction of vector mesons near the forward direction

$$\gamma p \rightarrow V^0 p; \quad V^0 = \varrho^0, \omega, \phi, \dots \tag{1}$$

belongs to the class of diffractive processes [1]

$$A + B \rightarrow A + X, \tag{2}$$

which are characterized by no quantum number (Q, I, S, B, C) exchange and by a seemingly constant, non-vanishing cross-section at high energies [2, 3]. Here, X may be either a resonance or a non-resonating multiparticle system. Known examples of such diffraction dissociation channels are [4, 5]

$$
\begin{aligned}
\pi^{\pm} p &\to (3\pi)^{\pm} p & [A_1] \\
\pi p &\to \pi (N\pi)^+ & [N^*(1470), N^*(1690)] \\
&\quad \pi (N\pi\pi)^+ & [N^*(1470), N^*(1690)] \\
K^{\pm} p &\to (K^{\pm}\pi^+\pi^-) p & [Q] \\
\gamma p &\to (\pi^+\pi^-\pi^0) p & [\omega] \\
&\quad (\pi^+\pi^-) p & [\varrho^0] \\
&\quad (K^+ K^-) p & [\phi] .
\end{aligned} \tag{3}
$$

These reactions are commonly thought of to proceed by pomeron ($\mathbb{P}$) exchange, as indicated in Fig. 1. Needless to say, the diffractive channels are going to be the most important single inelastic channels in the region of several hundred GeV, where the inelastic reactions involving quantum number exchange are expected to be strongly suppressed for their lower lying trajectories.

The classical empirical approach to such few body final state reactions is to extract a quasi-two-body final state from the data and to compare this sample with theoretical models (for the diffraction dis-

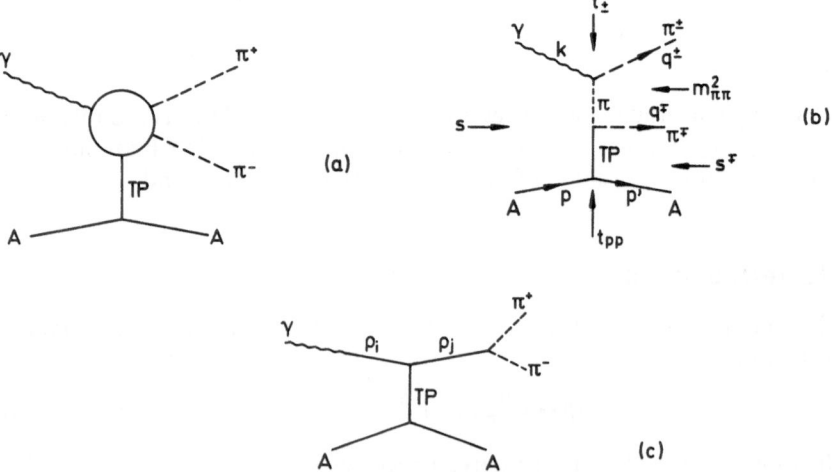

Fig. 1 a Example of a diffraction dissociation process [Eq. (2)], with $A = p$, $B = \gamma$, $X = \pi^+\pi^-$.
b Drell diagram for $\gamma p \to \pi^+\pi^- p$, showing our notation. c Vector meson dominance graph

sociation processes given in (3), we have indicated the most important observed resonances in brackets). This procedure works very nicely in the case of narrow resonances on top of a small background, like in ω or ϕ photoproduction [5], but it becomes problematical in the case of a large diffractively produced background as in the hadronic examples listed above, where the very resonance character of the A_1, Q and $N^*(1470)$ bumps is even in dispute [6].

The 2π photoproduction is an intermediate case [5]: the 2π mass spectrum is indeed strongly ϱ dominated, but the ϱ shape shows a dramatic deviation from the Breit-Wigner form [7]. This phenomenon has for long been a nuisance to anyone interested in ϱ physics, and various prescriptions have been published to extract the "ϱ" out of the data. Unfortunately, the resulting "ϱ" cross-sections depend on the recipe by which they were cooked up. But there is some interesting physics in this annoying phenomenon, and hence one should regard it as a tool to learn something about diffraction dissociation, following Harari's surmise that the best place to learn about hadron physics is an electron accelerator.

In this talk, I shall try to review the status of those models, that have been proposed with this motivation and, therefore, treat diffractive 2π photoproduction as a whole. (These models, of course, apply, mutatis mutandis, also for $K\bar{K}$ photoproduction, for which only meager data exist; 3π photoproduction must have frightened the theorists for its complexity, and maybe the experimental status was not such a challenge to them). I shall say very little about the vector meson dominance model in $\gamma p \to V p$ and nothing on vector meson production on nuclei and $\varrho^0 - \omega$ interference which shall be covered by other speakers during the Summer Institute.

Before I come to diffractive models for $\gamma p \to \pi^+ \pi^- p$ in Section 3, let me first give a brief account on the competing, i.e., non-diffractive contributions to $\gamma p \to V p$.

2. Non-Diffractive Contributions

2.1. Some $SU(3)$ Arguments about Couplings

In $\gamma p \to V^0 p$, the main competitor to $\mathbb{P}$ exchange is expected to be π^0 exchange. By $SU(3)$ arguments with ideal mixing, one finds the $\omega \pi \gamma$ coupling to be three times stronger than the $\varrho^0 \pi^0 \gamma$ coupling, and the $\phi \pi \gamma$ coupling zero [8]

$$\lambda_{\varrho^0 \pi \gamma} : \lambda_{\omega \pi \gamma} : \lambda_{\phi \pi \gamma} = 1 : 3 : 0 . \tag{4}$$

On the other hand, the $\mathbb{P}$ couplings can easily be related on the assumption that $\mathbb{P}$ is a $SU(3)$ singulet and γ a u spin scalar [9]:

$$\lambda_{\gamma\,\mathbb{P}\,\varrho^0} : \lambda_{\gamma\,\mathbb{P}\,\omega} : \lambda_{\gamma\,\mathbb{P}\,\phi} = 3 : 1 : -\sqrt{2} \,. \tag{5}$$

Although the last ratio of Eq. (5) is badly broken experimentally $(\sigma_\phi^{\text{Diff}}/\sigma_\varrho^{\text{Diff}} = 0.6\,\mu\text{b}/15\,\mu\text{b}!)$, which can be related by quark model arguments to the $SU(3)$ breaking observed in $\sigma_{\pi p}^{\text{tot}} \neq \sigma_{Kp}^{\text{tot}}$ [9] in the diffraction region, Eqs. (4) and (5) nevertheless teach us that ϱ^0 photoproduction has the largest diffractive contributions on one hand and the smallest π exchange contributions on the other, the inverse being true for ω photoproduction, while ϕ photoproduction has no π exchange at all and is thus the best reaction to study diffraction [10] if it were not for the minute cross-section.

2.2. Experimental Checks for $\mathbb{P}$ Exchange

There are essentially three pieces of evidence that hint at the dominating role of $\mathbb{P}$ exchange in $\gamma p \to V p$: constant cross-section, isospin zero and natural parity in the t channel (the latter being suggested by the idea of a universal pomeron).

a) Energy Dependence

Experimentally, ϱ^0 and ϕ photoproduction are fairly constant above 5 GeV, while ω photoproduction has a marked energy dependence [5, 7]. Writing its cross-section as an incoherent sum of $\mathbb{P}$ and π exchange

$$\sigma(E_\gamma^{\text{lab}}) = \sigma_{\mathbb{P}} + c/E_\gamma^{\text{lab}2} \,, \tag{6}$$

one can determine empirically the diffractive part, $\sigma_{\mathbb{P}}$, to be $\approx 2\,\mu\text{b}$ [in nice agreement with Eq. (5) and $\sigma_\varrho \approx 15\,\mu\text{b}$] and the constant c, that rules the strength of one pion exchange. In the framework of the absorption model, c can be related to the width $\Gamma_{\omega\to\pi\gamma}$ and one obtains this way [11]

$$\Gamma_{\omega\to\pi\gamma} \approx 1.2\,\text{MeV} \,, \tag{7}$$

which is a reasonable value.

b) t Channel Isospin

The t channel isospin can be studied by looking at the photoproduction on deuterons:

$$\gamma d \to d(np)\,\varrho^0 \,. \tag{8}$$

The ratio of the forward production cross-sections on deuterons and protons gives an indication for the presence of t channel isospin 1 and 0 exchange amplitudes, $A^{(1)}$ and $A^{(0)}$. In the impulse approximation, using closure and *neglecting* spin-flip amplitudes, one finds

$$R_{DH} \equiv \frac{d\sigma_D}{d\sigma_P}\bigg|_{0^\circ} = \frac{4|A^{(0)}|^2}{|A^{(1)} + A^{(0)}|^2} .$$ (9)

With Glauber corrections one would expect R_{DH} to be about 3.6 in case of pure $\mathbb{P}$ exchange. Bubble chamber data yield a 10–20% lower value [12]. The situation can best be assessed by looking at Fig. 2 that

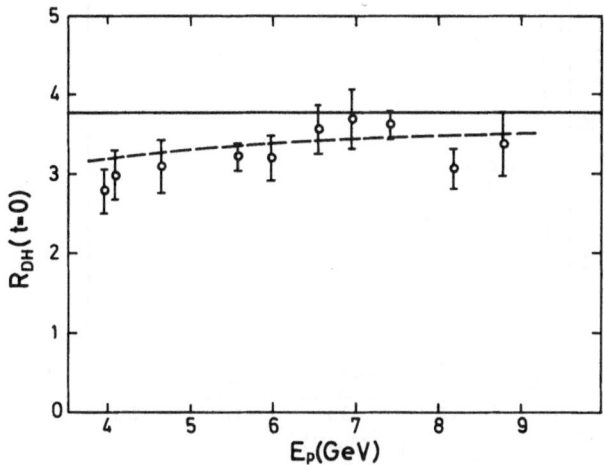

Fig. 2. The ratio of deuteron to hydrogen cross-sections [Eq. (9)], taken from Ref. [13], as a function of E_γ^{lab}

shows the results of the most recent analysis of the Cornell group on their counter data [13]. Since the determination of the forward production cross-section depends rather sensitively on the method of data analysis and since one has to expect some 10–20% background contamination of non 2π events in the counter experiments [5], I would see no compelling evidence for the presence of $I_t = 1$ amplitudes above 4 to 5 GeV from this data. On the other hand, a value of $|A^{(1)}|/|A^{(0)}|$ of the order of $\sim 10\%$ cannot be definitely excluded.

A direct way of looking at $I_t = 1$ exchange is, of course, to study charged ϱ photoproduction

$$\gamma n \to \varrho^- p .$$ (10)

Fig. 3 shows the experimental result [14], whose order of magnitude is in agreement with one-pion exchange, with a width of $\Gamma_{\varrho^0 \to \pi^0 \gamma} = 0.13$ MeV.

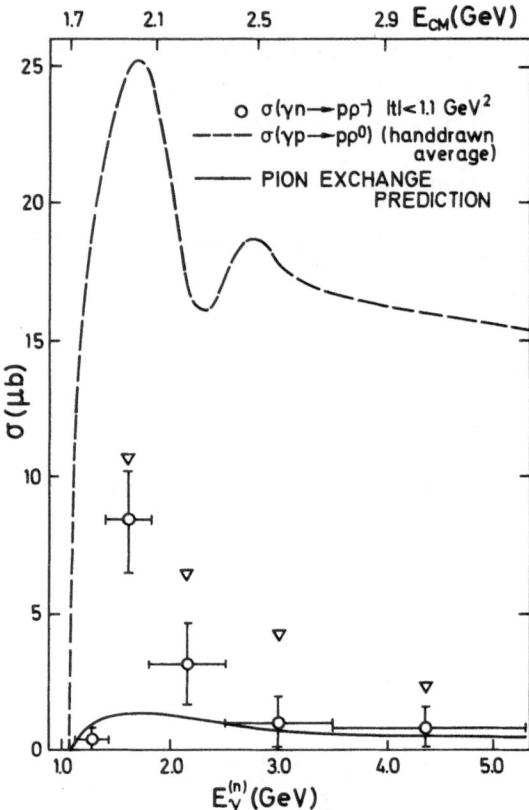

Fig. 3. Cross-section for $\gamma n \to p\varrho$ for $|t_{pp}| < 1.1\ \text{GeV}^2$ as a function of E_γ^{lab}, taken from Ref. [14]. The solid curve is the prediction of π exchange with $\Gamma_{\varrho\pi\gamma} = 0.13$ MeV. The triangles give an upper limit of the cross-section for all $|t_{pp}|$

c) t Channel Normality

The t channel normality, defined as $N_{\text{ex}} = P \cdot \tau$, with P, τ the parity and signature of the exchanged trajectory, can be readily determined with the help of polarized photons. Given a definite normality N_{ex} in the t channel, the s channel helicity amplitudes obey the symmetry relation [15]

$$T^{(s)}_{-\lambda_V \lambda_{N'},\, -\lambda_\gamma \lambda_N} = - N_{\text{ex}}(-)^{\lambda_V}\, T^{(s)}_{\lambda_V \lambda_{N'},\, \lambda_\gamma \lambda_N} \tag{11}$$

to leading order in s. Exploiting this relation, one can construct combinations of density matrix elements that correspond to a pure normality N_{ex} [16]. For that purpose, the photon density matrix is parametrized by the Stokes parameters P_1, P_2, P_3 as

$$\varrho^\gamma = \tfrac{1}{2}(I + \boldsymbol{P}\boldsymbol{\sigma}), \tag{12}$$

where I is a 2×2 unit matrix and σ are the usual Pauli matrices. The vector meson spin density matrix, defined by

$$\varrho^V = T^{(s)} \varrho^\gamma T^{(s)+} \tag{13}$$

can now be expanded in the basis

$$(\varrho^0, \varrho) = \tfrac{1}{2} T^{(s)} (I, \sigma) T^{(s)+} \tag{14}$$

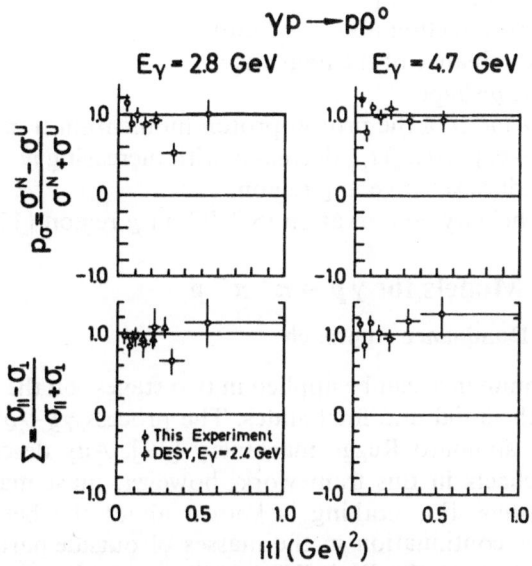

Fig. 4. P_σ and Σ in the reaction $\gamma p \to p \varrho^0$ for two photon energies. $P_\sigma = \Sigma = 1$ means pure natural parity in the t channel. Figure taken from Ref. [17]

with all nucleon helicities summed over. ϱ^0, ϱ contain the full information obtainable from experiments without nucleon polarization. It turns out, that they contain no interference terms between natural ($N_{ex} = +1$) and unnatural ($N_{ex} = -1$) parity exchange contributions.

$$\varrho^j = \varrho^j_{N_{ex} = +1} + \varrho^j_{N_{ex} = -1}; \quad j = 0, ..., 3. \tag{15}$$

The latter can be obtained through relations like

$$\varrho^{1 N_{ex}}_{\lambda \lambda'} = \tfrac{1}{2} (\varrho^1_{\lambda \lambda'} - N_{ex}(-)^\lambda \varrho^0_{-\lambda \lambda'}) \\ \varrho^{0 N_{ex}}_{\lambda \lambda'} = \tfrac{1}{2} (\varrho^0_{\lambda \lambda'} - N_{ex}(-)^\lambda \varrho^1_{-\lambda \lambda'}). \tag{16}$$

Furthermore, the parity asymmetry P_σ can be written in terms of ϱ^1:

$$P_\sigma \equiv \frac{\sigma^{nat} - \sigma^{unnat}}{\sigma^{nat} + \sigma^{unnat}} = 2\varrho^1_{1-1} - \varrho^1_{00}. \tag{17}$$

Fig. 4 shows the result of the 4.7 GeV SLAC-Tufts-Berkeley HBC experiment on $\gamma p \to \varrho p$ [17]. Within small error, P_σ is equal to one, i.e.,

$N_{ex} = +1$. For ω production, on the other hand, one has a sizeable energy dependent $N_{ex} = -1$ component [17], in agreement with one-pion exchange, whereas σ^{nat} looks fairly constant between 2.8 and 4.7 GeV. So we have good evidence that 2π photoproduction is a process with a large diffractive component. Before I proceed to discuss more detailed models on $\gamma p \to 2\pi p$, let me give the complete list of experimental features of this process [5, 7]:

   a) constant cross-section above ~ 5 GeV,

   b) strong ϱ dominance of the final state,

   c) asymmetric ϱ shape,

   d) the slope $A(m_{\pi\pi})$ of the proton-proton momentum transfer distribution, $d\sigma/dt_{pp} \sim \exp A(m_{\pi\pi}) t_{pp}$ decreases with increasing $m_{\pi\pi}$,

   e) natural parity exchange in ϱ region,

   f) s channel helicity conservation (SCHC) in ϱ region [17, 18].

3. Diffractive Models for $\gamma p \to \pi^+ \pi^- p$

3.1. The Vector Dominance Approach

Vector meson dominance can be applied in two stages: on the final state only and on both initial and final states. The process $\gamma p \to \varrho^0 p$ can be treated with the standard Regge machinery [19]. Any description of $m_{\pi\pi}$ dependent effects in this framework, however, must make use of ad hoc assumptions. For nothing is known about the behaviour of amplitudes under continuation in the masses of outside particles. E.g., a $1/m_{\pi\pi}^2$ modification to the Breit-Wigner shape can be accomplished by choosing the kinematic singularity free t channel helicity amplitudes $m_{\pi\pi}$ independent [20]. Apart from this, all the other observed features of our process can be naturally incorporated into the Regge framework [21].

The second stage vector meson dominance relates our process to elastic hadronic scattering, $\varrho p \to \varrho p$. As already mentioned in the Introduction, this sort of approach has nothing to say about $m_{\pi\pi}$ dependent effects, the famous Ross-Stodolsky factor $1/m_{\pi\pi}^2$ [22] being again due to an ad hoc smoothness assumption. On the other hand, second stage vector meson dominance has led to a number of interesting relations between hadronic and electromagnetic phenomena that shall be discussed in the talks of Professor Gottfried and Dr. Schildknecht during this symposium.

3.2. The Multiperipheral Model

This model is just opposite to the vector meson dominance idea: the $\pi^+ \pi^-$ pair is produced in a continuum state, by dissociation of the incoming γ into $\pi + \chi$ and subsequent diffractive scattering of χp into πp, χ being either π or A_1. The $\pi\pi$ system having $G = 1$, it can only be diffractively produced by the isovector photon, leading to isovector 2π final states with odd angular momentum. Drell originally assumed only intermediate pions [23] (see Fig. 1 b). Naively, one would write for the two possible Drell diagrams, neglecting the unobserved nucleon spin

$$T = e\{\varepsilon q^+ \, T_-/k q^+ - \varepsilon q^- \, T_+/k q^-\} \tag{18}$$

with

$$T_\pm = i s_\pm \sigma^{tot}_{\pi^\pm p} \exp(\tfrac{1}{2} B t_{pp}) f(t_\pm)$$

$$s = m^2_{\pi\pi} = (q^+ + q^-)^2, \qquad s_\pm = (p' + q^\pm)^2$$

$$t_\pm = (q^\pm - k)^2, \qquad \bar{s} = (k + p)^2$$

$$t_{pp} = (p' - p)^2 \, .$$

For notations see Fig. 1 b. Apart from the factor $f(t_\pm)$, which describes the possible off-shell dependence of the intermediate pions, $T_\pm$ is the amplitude of diffractive πN scattering, i.e., $B \approx 8 \, \text{GeV}^{-2}$.

The expression Eq. (18) for T is only gauge invariant on the pion pole; as one goes away from the pole, one has to write some gauge invariant extension, which, unfortunately, is rather arbitrary. For instance, one might replace [24]

$$\varepsilon \cdot q^\pm \to \varepsilon \cdot q^\pm - (k \cdot q^\pm/k \cdot a) \, a \cdot \varepsilon \tag{19}$$

with a any of the four-vectors appearing in the problem; e.g., one might choose $a = p + p' = P^+$ which leads to an unphysical pole far off at $k P^+ \sim \bar{s} - \bar{u} = 0$ [25]. Or one might split off a gauge invariant part and make the rest gauge invariant by extension:

$$\begin{aligned} T \to &e\{\varepsilon q^-/k q^- - \varepsilon q^+/k q^+\} \tfrac{1}{2}(T_+ + T_-) \\ &+ e\{\varepsilon q^-/k q^- + \varepsilon q^+/k q^+ - \varepsilon a/k a - \varepsilon a'/k a'\} \tfrac{1}{2}(T_+ - T_-) \, . \end{aligned} \tag{20}$$

This was the procedure of Kramer et al. [26], who chose $a = p$, $a' = p'$, thus obtaining nucleon pole terms, which are hard to interpret in terms of diffraction. Another prescription is to set the term proportional to $T_+ - T_-$ equal to zero. With $\tfrac{1}{2}(s_+ + s_-) \sim \bar{s}$, this is equivalent to a Pomeron of effective spin 0^+, which has been assumed in dual diffractive dissociation models [27, 28]. It might be comforting to realize that all these recipes coincide for forward production, if one chooses the Coulomb gauge either in the over-all centre-of-mass system or in the 2π rest system. For the general kinematical situation, however, one should keep in mind that this ill-definedness of the Drell amplitude has great bearing on the 2π decay distribution. We, therefore, expect the Drell model to have predictive power only on the more global properties of the process such as $d\sigma/dt_{pp}$, $d\sigma/dm_{\pi\pi}$.

3.3. The Interference Model

The interference model [29] writes the 2π photoproduction amplitude as a sum of a directly produced Breit-Wigner shaped ϱ plus the Drell term and thus successfully describes the ϱ asymmetry by the interference of the real parts of the two terms (see Fig. 5). It also predicts the mass dependence of the t_{pp} slope, but not the ϱ polarization and the absolute value of the cross-section.

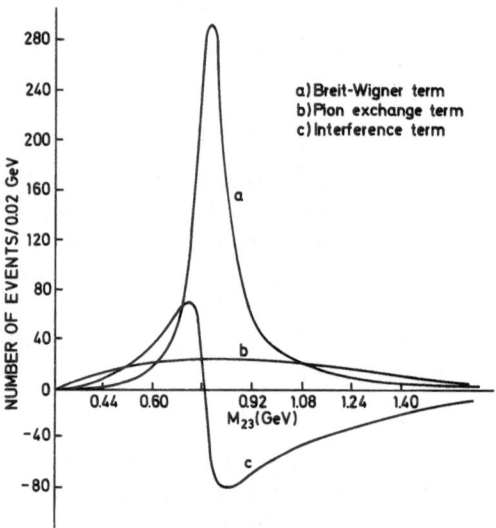

Fig. 5. The three contributions to $d\sigma/dm_{\pi\pi}$ in the interference model. Figure taken from Ref. [25]

It has often been objected that the interference model implies double counting. This is not so obvious in the framework of finite-energy-sum rule duality, which only connects the imaginary parts of the direct channel with the imaginary parts of the crossed channels. The success of the interference model, however, is mainly a *real* part effect. Yet, it is not clear, how to approximate the real part. How much of the real part of the t channel Regge exchange amplitude should be added to the s channel resonance, after the resonance has been given a real part à la Breit-Wigner [30]?

A more serious objection to the interference model is that it does not reproduce the $\pi-\pi$ p wave scattering phase δ as predicted by the Watson theorem, which we would expect to hold in $\gamma\,\mathbb{P}\rightarrow\pi\pi$ – although it cannot be proved here rigorously as in genuine two-body photoproduction processes. It has been suggested to achieve the expected phase by applying a factor $\cos\delta\, e^{i\delta}$ to the p wave part of the Drell diagram before adding it to the ϱ amplitude [24, 31]. For small δ, this factor corresponds to final state interaction. For large δ, it can hardly be justified this way. In fact, one would expect the final state interaction to bring about an enhancement in the ϱ region. This brings us directly to the final state interaction model.

3.4. Final State Interaction Model

Kramer and Uretsky [26] carried out the idea that the ϱ photoproduction can be visualized as a final state interaction effect in the p wave part of the Drell amplitude. Symbolically

$$T_{\gamma p \to \varrho p} = T_{\text{Drell}}^{p\text{-wave}} \times \text{final state interaction.} \tag{21}$$

In their original version, they chose an enhancement factor E to account for final state interaction

$$E(m_{\pi\pi}) = \lambda\, [m_\varrho^2 - m_{\pi\pi}^2 - i m_\varrho \Gamma_\varrho^{\text{tot}}]^{-1}. \tag{22}$$

The normalization λ remains an open constant. In a more refined version of this model, whose details we cannot cover here, Kramer and Quinn [26] set up and approximately solved the N/D problem for the $2\pi\, p$ wave. Again, the over-all normalization is undetermined [32]. It is related to the unknown behaviour of the p wave amplitude for large $m_{\pi\pi}$. Their result is essentially

$$\frac{d^2\sigma}{dt_{pp}\,dm_{\pi\pi}^2}\bigg|_{p\text{-wave}} \tag{23}$$
$$= \frac{\pi m_\varrho \Gamma_\varrho(m_{\pi\pi})}{(m_\varrho^2 - m_{\pi\pi}^2)^2 + m_\varrho^2 \Gamma_\varrho^2(m_{\pi\pi})} \cdot \left(\frac{m_\varrho^2 - t_{pp}}{m_{\pi\pi}^2 - t_{pp}}\right)^2 \frac{4\alpha^2 \lambda^2}{(g_{\varrho\pi\pi}^2/4\pi)} \frac{d\sigma_{\pi p}}{dt_{pp}}$$

with $\Gamma_\varrho(m_{\pi\pi})$ the Jackson p wave width.

Fig. 6 to Fig. 8 show that this formulation describes very well the shape of $d\sigma/dm_{\pi\pi}$ and $d^2\sigma/dt_{pp}\,dm_{\pi\pi}$. This is essentially due to the factor $1/(m_{\pi\pi}^2 - t_{pp})^2$, whose origin is the pion propagator which can be written as

$$(t_\pm - \mu^2)^{-1} = (2q^\pm k)^{-1} \sim (m_{\pi\pi}^2 - t_{pp})^{-1} \cdot (1 \pm b\cos\theta_{\gamma\pi})^{-1} \tag{24}$$

in the 2π rest system, with b a smooth function. This factor is a good old friend by now: except for the t_{pp} term, it is the Ross-Stodolsky factor. The appearance of t_{pp} is mainly responsible for the successful description

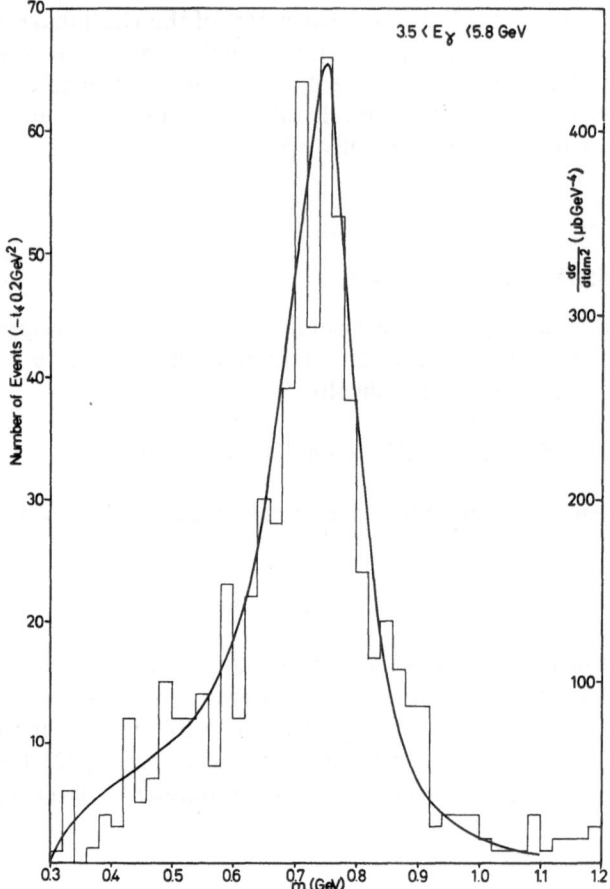

Fig. 6. Prediction of final-state interaction model on $d\sigma/dm_{\pi\pi}^2$. Data from Ref. [41]. Figure taken from Ref. [26b]

of the slope variation with mass, $A(m_{\pi\pi})$, (see Fig. 9). s channel helicity conservation, on the other hand, is not predicted by this model. It can be built in, however.

3.5. Veneziano Approach

One might try to extend the idea of duality [33] to diffraction dissociation by a B_4 description [34] of the bubble $\gamma + \mathbb{P} \to \pi^+ \pi^-$ [28] (see Fig. 1a). With the simplest choice of $J^p = 0^+$ for the pomeron, the amplitude for $\gamma \mathbb{P} \to \pi\pi$ is

$$A = \{(\varepsilon Q^+)(kQ^-) - (\varepsilon Q^-)(kQ^+)\} V(s, t, u) \tag{25}$$

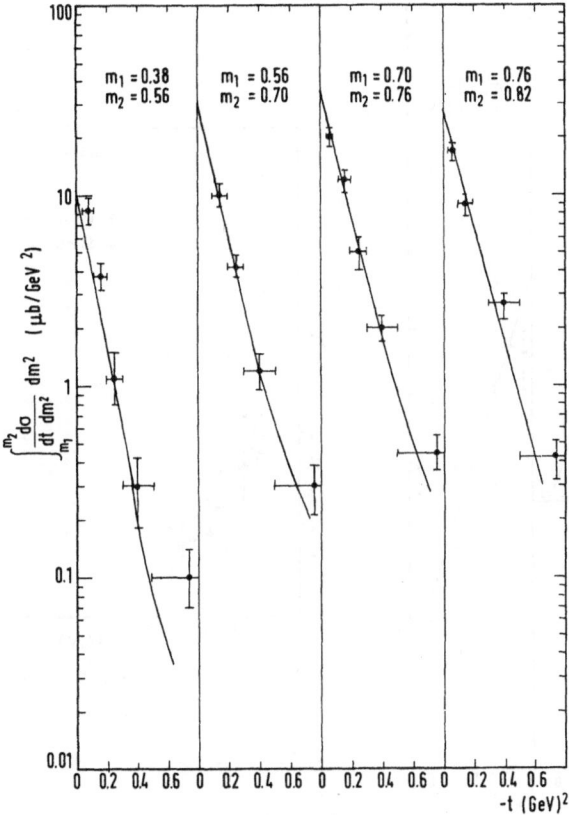

Fig. 7. Prediction of final-state interaction model on $d\sigma/dt_{pp}$, in various $m_{\pi\pi}$ bins. Figure taken from Ref. [26 b]

with $Q^{\pm} = q^{+} \pm q^{-}, s = m_{\pi\pi}^2, t = t^{+}, u = t^{-}$. Insisting on parent-parent duality, we write a Veneziano form for the invariant amplitude V

$$V(s, t, u) = [\beta/(\alpha_t + \alpha_u)] \{B(1 - \alpha_\varrho, -\alpha_t) + (t \rightarrow u) + \eta B(-\alpha_t, -\alpha_u)\}, \quad (26)$$
$$B(x, y) = \Gamma(x) \Gamma(y)/\Gamma(x + y)$$

with α_t, α_u the pion trajectories in the t, u channels and α_ϱ the direct channel ϱ trajectory (all of slope 1 GeV^{-2}) with an imaginary part linear in s, reproducing the ϱ width. This form has the correct pion poles and the correct Regge behaviour. The amplitude for the full process $\gamma p \rightarrow \pi^{+} \pi^{-} p$ is then written

$$T = i \bar{s} (\exp \tfrac{1}{2} B t_{pp}) A (s, t, u) \quad (27)$$

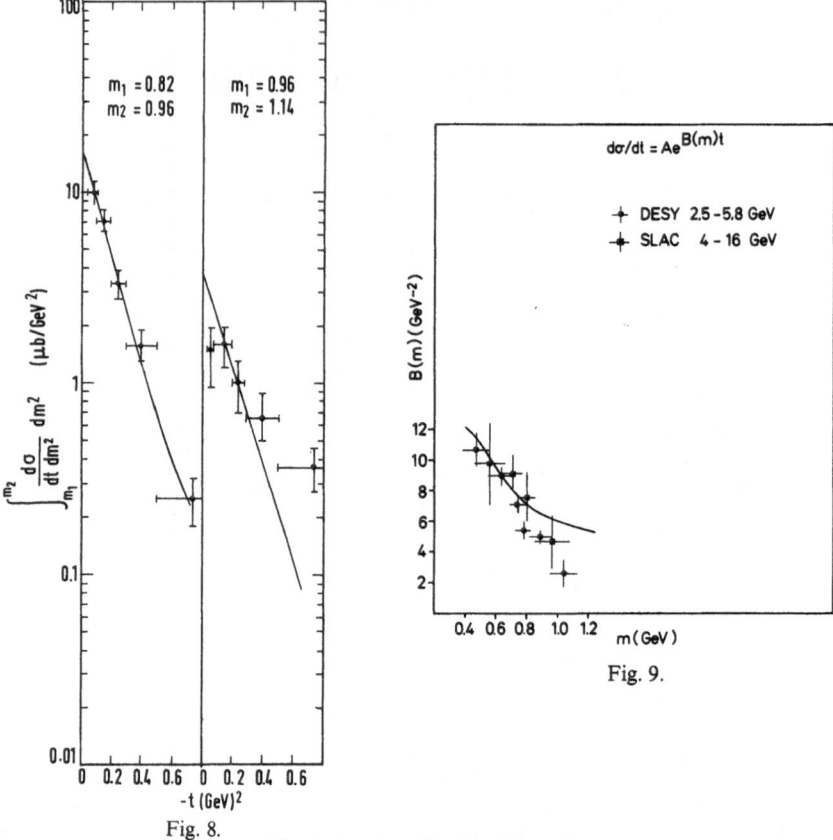

Fig. 8.

Fig. 8. Same as Fig. 7, continued

Fig. 9. Variation of the t_{pp} slope with $m_{\pi\pi}$ in the final-state interaction model. Figure taken from Ref. [26b]

with $(\exp\frac{1}{2}Bt_{pp})$ coming from the $(pp\mathbb{P})$ form factor, i.e., $B = 5\ \text{GeV}^{-2}$. There are two parameters in this model, β and η. It turns out that $\eta = 0$ is empirically the best value, which implies strong $\pi - A_1$ exchange degeneracy (there is no theoretical reason, however, for such exchange degeneracy in this model). If β is fitted to the total reaction cross-section $\sigma(\gamma p \to \pi^+ \pi^- p)$, the $m_{\pi\pi}$ distribution and differential cross-section are nicely reproduced (see Figs. 10, 11). In particular, $A(m_{\pi\pi})$ decreases from a large value to $5\ \text{GeV}^{-2}$ for large $m_{\pi\pi}$, with a minimum due to the resonances in between, as can be seen in Fig. 12. Such structure is also seen in the SLAC data, which also yield a value of $5\ \text{GeV}^{-2}$ for the t_{pp} slope of ϱ production [17], in agreement with the present model (see Fig. 12b).

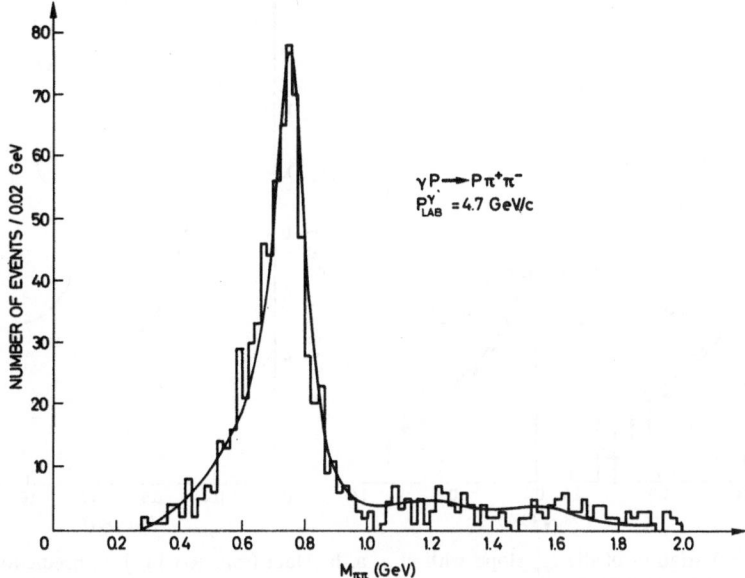

Fig. 10. The predicted $d\sigma/dm_{\pi\pi}$ distribution of the dual model of Ref. [28]

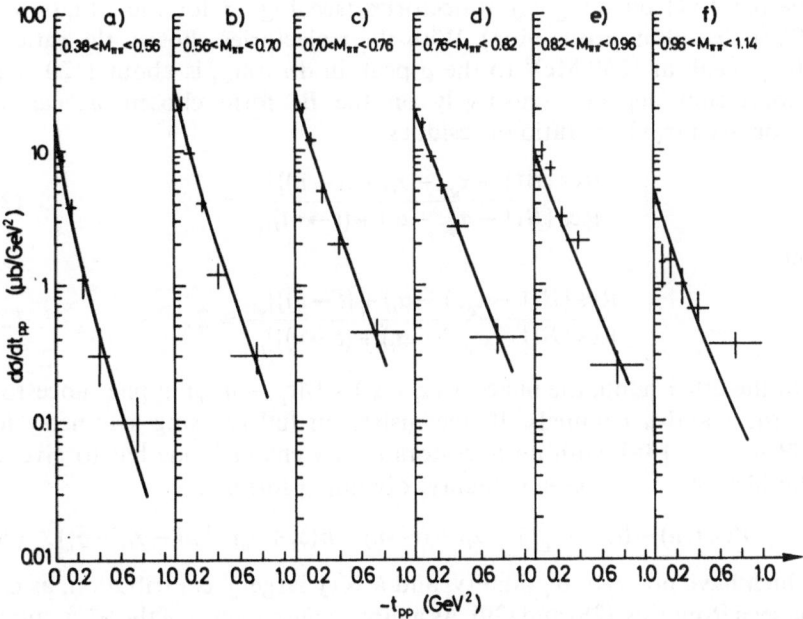

Fig. 11. $d\sigma/dt_{pp}$, as predicted by the dual model of Ref. [28]. Data from Ref. [41]

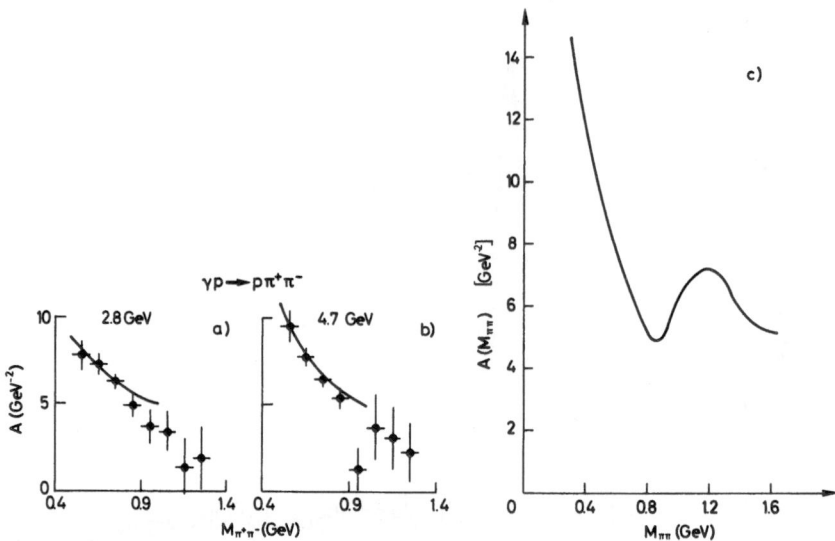

Fig. 12. Variation of the t_{pp} slope with $m_{\pi\pi}$: a, b: Data from Ref. [17], c: prediction of dual model of Ref. [28]

As is well known, the Veneziano model predicts higher lying vector mesons [34] on daughter trajectories (see Fig. 13 for the situation in diffractive photoproduction). With the trajectories chosen, the ratio of the ϱ' peak at 1250 MeV to the ϱ peak in $d\sigma/dm_{\pi\pi}$ is about 1:20. This suppression depends sensitively on the B_4 form chosen, as can be recognized from the ratio of residues

$$\frac{\text{Res}\{B(1-\alpha_\varrho,-\alpha_t)+(t\to u)\}|_{\varrho'}}{\text{Res}\{B(1-\alpha_\varrho,-\alpha_t)+(t\to u)\}|_\varrho}=\tfrac{1}{4},\qquad(28)$$

but

$$\frac{\text{Res}\{B(1-\alpha_\varrho,1-\alpha_t)+(t\to u)\}|_{\varrho'}}{\text{Res}\{B(1-\alpha_\varrho,1-\alpha_t)+(t\to u)\}|_\varrho}=2.\qquad(29)$$

On the other hand, the factor $1/(\alpha_t+\alpha_u)\sim 1/(t_{pp}-m_{\pi\pi}^2)$ implies ancestors in the t and u channels. If one insists on full crossing symmetry for $\gamma\mathbb{P}\to\pi^+\pi^-$ [35], with no ancestors in any channel, one has to give up the idea of parent-parent duality, obtaining forms like

$$V(s,t,u)\sim B(1-\alpha_\varrho,1-\alpha_t)+(t\to u)+\eta\,(\alpha_t+\alpha_u)^{-1}B(-\alpha_t,-\alpha_u),\quad(30)$$

which have no π–ϱ B_4 duality, and a very large ϱ' contribution, as can be seen from Eqs. (28) and (29). As a direct consequence of the 0^+ assumption for the pomeron, the model predicts t channel helicity conservation

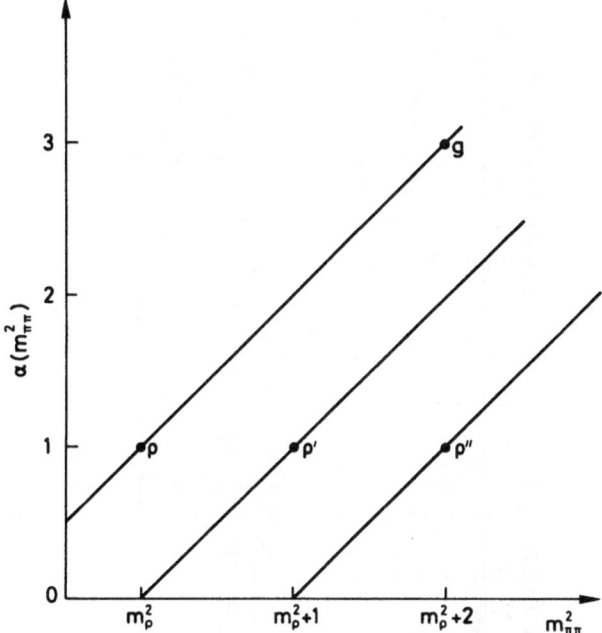

Fig. 13. The 2π spectrum of particles predicted to be diffraction produced in the dual model of Ref. [28] for the reaction $\gamma p \to \pi^+ \pi^- p$

for the ϱ pole [18], which is meanwhile ruled out by experiment [17]. Outside the ϱ band, the 2π decay distribution agrees nicely with the model (see Fig. 14). The significance of the data for $m_{\pi\pi}$ above 1 GeV is not clear, however, since one expects here reflections from $\varDelta$ production.

It is interesting to note that the interference model is a good approximation to the B_4 model in the vicinity of ϱ:

$$B(1-\alpha_{\varrho'}-\alpha_t) \approx (1-\alpha_\varrho)^{-1}+(-\alpha_t)^{-1}, \tag{31}$$

as can be seen from Fig. 15. In that figure the ϱ mass has been taken at 765 MeV. The resulting peak is at 745 MeV.

This simple model was modified by Dewey and Humpert [25] to include the double Regge limit, absolute normalization at the pion pole and to reproduce SCHC at the ϱ pole in the limit t_{pp} fixed, $\bar{s} \to \infty$, giving up the 0^+ assumption for the pomeron. Their ansatz is:

$$T = i \cdot 2e\sigma_{\pi N}^{\text{tot}} (\exp{\tfrac{1}{2}Bt_{pp}}) \, [Pq^+ \{q^- \varepsilon - (q^- k)(P\varepsilon)/Pk\} \, B(1-\alpha_\varrho, \, -\alpha_t)$$
$$-Pq^- \{q^+ \varepsilon - (q^+ k)(P\varepsilon)/Pk\} \, B(1-\alpha_\varrho, \, -\alpha_u)], \tag{32}$$

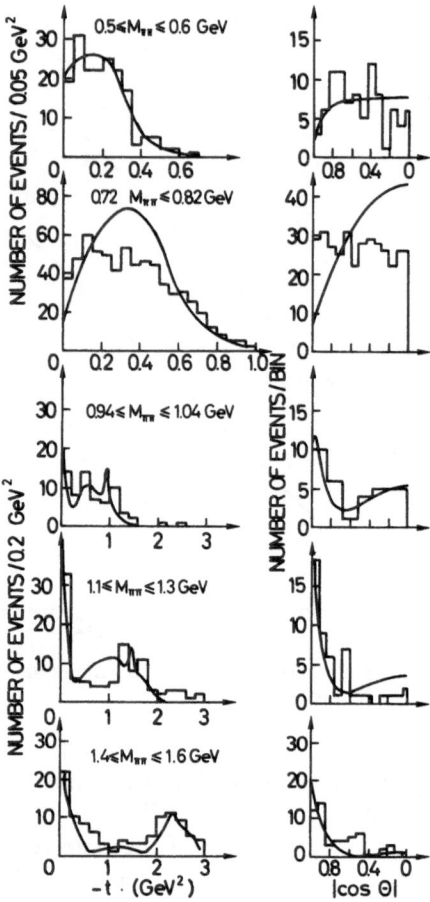

Fig. 14. The π decay distributions, plotted versus t_+ and the cosine of the Jackson angle for various $m_{\pi\pi}$ intervals. Data from Ref. [41], theoretical curves predicted by the dual model of Ref. [28]

with $P = p + p'$. In the forward direction, this form coincides on the ϱ pole with the previous ansatz, so the main features of its predictions are the same as in the previous model (see Fig. 16). Again, strong πA_1 exchange degeneracy is implied. The decay angular distributions in the ϱ region are now very nicely reproduced at 4.7 GeV (see Fig. 17). The details of the ϱ' prediction, however, are very doubtful, since the ansatz Eq. (32) has ancestors in the direct channel, due to the Pq^+, Pq^- factors, which are first order polynomials in the cosine of the $\pi^\pm$ direction. Hence the ancestors show up at odd spins on a trajectory one unit higher than the ϱ trajectory. There is a way out, of course, as indicated

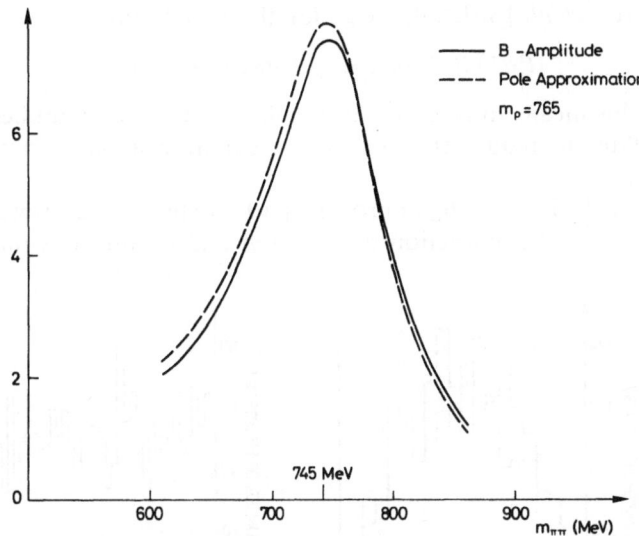

Fig. 15. Comparison between the prediction of the B_4 function and its approximate pole form [Eq. (32)] on $d\sigma/dm_{\pi\pi}$

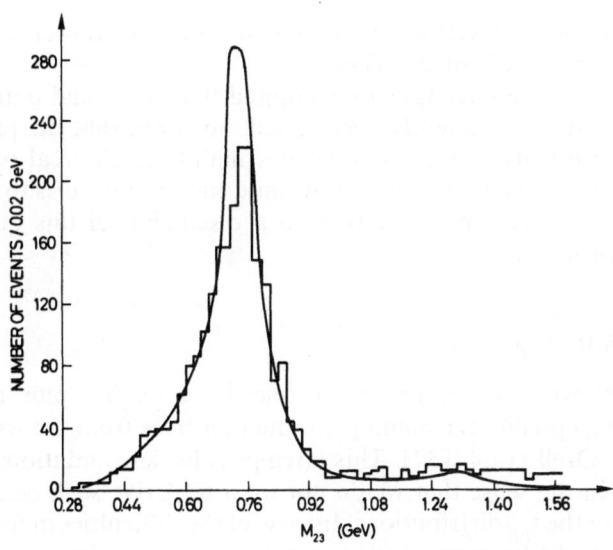

Fig. 16. $d\sigma/dm_{\pi\pi}$, as predicted by the dual model of Ref. [25]. Data taken from Ref. [17]

by Dorren *et al.* [36]: take (e.g., for the $s-t$ term):

$$(Pq^+)\,B(2-\alpha_\varrho,\,-\alpha_t)+b\bar{s}\,B(1-\alpha_\varrho,\,1-\alpha_t)\,. \tag{33}$$

Again, this means to give up parent-parent duality and implies the need of satellites to reduce the large ϱ' contribution arising from the form Eq. (33).

In conclusion, the B_4 approach is a first step to incorporate duality ideas into $\gamma \to 2\pi$ diffraction dissociation and is still on a rather crude

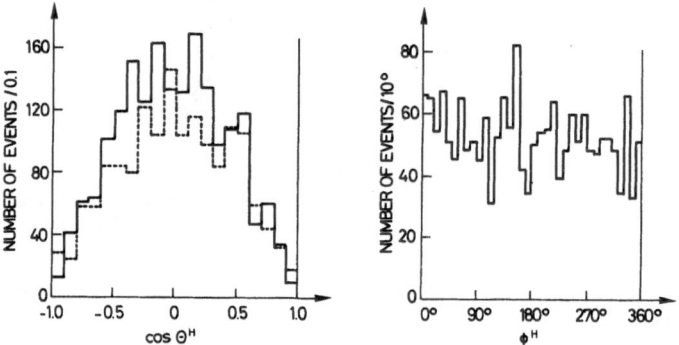

Fig. 17. The prediction of the dual model of Ref. [25] (full histogram) against the experimental results of Ref. [17] (dotted histogram) on the π angular distribution in the helicity frame ($|t_{pp}| < 0.4 \text{ GeV}^2$, $0.65 < m_{\pi\pi} < 0.85 \text{ GeV}$)

phenomenological level: in order to gain predictive power, one has so far to live with ancestor problems.

Up to now, we have taken for granted that the π and ϱ are dual to each other. It has frequently been stated, however, that the pion might fall outside duality, since it contributes mainly to the real part of the amplitude which is not connected by finite energy sum rules to the direct channel resonances. We now turn to a discussion of this question in 2π photoproduction.

3.6. F.E.S.R. Approach

In the following, we want to use semi-local duality plus resonance saturation to predict the ϱ and ϱ' production rates from the well-known size of the Drell graph [37]. This attempt is far less ambitious than the B_4 model in the sense that we do not want to derive such details as the ϱ shape and the t_{pp} distributions. In view of the difficulties in formulating the Drell model for non-forward production, we restrict our considerations to exact forward production.

We start with the "amplitude" for $\gamma\,\mathbb{P}\to\pi^+\pi^-$

$$T_4 = [(\varepsilon q^+)(kq^-) - (\varepsilon q^-)(kq^+)]\,A(s,t,u)\,, \tag{34}$$

that is based on the (fairly reasonable) assumption that the pomeron behaves like 0^+ for forward production (see the discussion in 3.2). The π exchange Born term reads

$$A_{\text{Bom}} = -4\,e g_{\mathbb{P}\pi\pi}/\{(t-\mu^2)(u-\mu^2)\}\,. \tag{35}$$

The Reggeized pion exchange is taken to be

$$R_\pm = -\pi e\,\frac{1+\exp[-i\pi\alpha_\pi(t_\pm)]}{\sin\pi\alpha_\pi(t_\pm)}\,\tfrac{1}{2}(\alpha_\pi(t_\pm)+2)\,(m_{\pi\pi}^2/v_0)^{\alpha_\pi(t_\pm)} \tag{36}$$

with $s,t,u = m_{\pi\pi}^2, t_+, t_-$. Then, the zeroth moment F.E.S.R. at fixed t yields a trivial relation between the t and u channel pion poles:

$$\tfrac{1}{2}\int_{-N}^{+N} dv\left\{\frac{2\pi e g_{\mathbb{P}\pi\pi}}{t-\mu^2}\,\delta(v-v_{\text{Bom}}) + \text{Im}\,A_{\text{non Bom}}(v,t)\right\} \tag{37}$$

$$= \frac{\pi e g_{\mathbb{P}\pi\pi}}{\alpha_\pi(t)}\,\tfrac{1}{2}(\alpha_\pi+2)\,(N/v_0)^{\alpha_\pi(t)}$$

with $v = \tfrac{1}{2}(s-u)$, which equals $v_{\text{Bom}} = \tfrac{1}{2}(s-\mu^2)$ at the u pole. A connection between the ϱ and π couplings can be obtained from the first moment sum rule

$$\int_{N_1}^{N_2} dv\left(\frac{v-v_{\text{Bom}}}{v_0}\right)\text{Im}\,A(v,t) = \frac{\pi e g_{\mathbb{P}\pi\pi}}{\alpha_\pi+1}\,\alpha_\pi'\tfrac{1}{2}(\alpha_\pi+2)\,(v/v_0)^{\alpha_\pi+1}|_{N_1}^{N_2}\,. \tag{38}$$

Saturating the left-hand side by ϱ, one finds at $t=\mu^2$

$$m_\varrho^2 g_{\gamma\varrho\mathbb{P}}\cdot g_{\varrho\pi\pi} = e g_{\mathbb{P}\pi\pi}\alpha_\pi'(N_2-N_1)\,. \tag{39}$$

After replacing $g_{\mathbb{P}\pi\pi}\to i\bar{s}\,\sigma_{\pi p}^{\text{tot}}$, one deduces

$$d\sigma/dt_{pp}|_{0^\circ} = \frac{q_\pi^3 e^2(\sigma_{\pi p}^{\text{tot}})^2(N_2-N_1)^2\alpha_\pi'^2}{384\,\pi^2 m_\varrho^2\,\Gamma_\varrho}\,. \tag{40}$$

The integration should cover the resonance spacing. If the next higher resonance is assumed to be at $m_\varrho' = 1.5$ GeV, as suggested by the nuclear experiments described below [38, 39], and α_π' is set equal to 1 GeV^{-2}, one has $\alpha_\pi'(N_2-N_1) = 1.4$ and Eq. (40) yields

$$d\sigma/dt_{pp}|_{0^\circ} \approx 50\,\mu\text{b/GeV}^2\,, \tag{41}$$

which is in fair agreement with the experimental value of $80-140\,\mu\text{b/GeV}^2$ [5]. Thus, in contrast to the wide-spread belief, the pion is a perfectly acceptable dual partner in the F.E.S.R. sense to the direct channel ϱ.

3.7. Why is ϱ' Photoproduction Suppressed?

A number of counter experiments of symmetric dipion production on nuclei have been performed to search for the photoproduction of high mass vector mesons in the coherence region [38, 39]. It has been found that $d\sigma/dm_{\pi\pi}$ drops by two orders of magnitude between m_ϱ and $m_{\pi\pi} = 1.5$ GeV, with a broad shoulder between 1.3 and 1.8 GeV, which might or might not be the ϱ' (or the g??).

In the following we want to show that such a dramatic suppression of the ϱ' (with $m_{\varrho'} = 1500$ MeV) production rate is to be expected on the basis of the Reggeized Drell mechanism and F.E.S.R. The argument goes as follows: the Reggeized Drell amplitude for $\gamma p \to \pi^+ \pi^- p$, in forward direction, is given by

$$T = e\{(\varepsilon q^-) R_- T_+ - (\varepsilon q^+) R_+ T_-\} \tag{42}$$

with $T_\pm$, $R_\pm$ defined by Eqs. (18) and (36), respectively. It contains only one open parameter, v_0. Choosing $v_0 = 0.7$ GeV^2, and normalizing the SLAC data [38], measured on Be, to the HBC data, one obtains a good average description of the experimental curve by the theoretical curve, calculated from Eq. (42) and

$$\frac{d\sigma}{dt_{pp}\,dm_{\pi\pi}\,d\Omega} = \frac{q_\pi}{256\,\pi^4(\bar{s}-m_p^2)^2}\,\tfrac{1}{2}\sum_{\lambda_\gamma}|T|^2. \tag{43}$$

Ω is the solid angle of the π direction in the 2π rest system (here $\theta = 90°$!). See Fig. 18.

Since the ϱ' is a broad object, $\Gamma_{\varrho'} \approx 300$ MeV, we would not expect it to protrude much over its background. This is also the answer from a semiquantitative application of F.E.S.R. to the ϱ' region, which we define to be the region of the shoulder mentioned above ($m_{\varrho'} \approx 1500$ MeV). The method is clearly less reliable here than in the case of the ϱ region, because one has to project out the p wave contributions of the Drell amplitude before one can relate the ϱ' residue to the π exchange. The final answer is [37]

$$\frac{d\sigma}{dt_{pp}\,dm_{\pi\pi}\,d\Omega}\Bigg|_{\substack{m_{\pi\pi}=1.5\ \text{GeV} \\ t_{pp}\approx 0 \\ \theta=90°}} = \frac{0.015}{[\Gamma_{\varrho'}^{\text{tot}}\ \text{in GeV}]^2}\ \mu\text{b/GeV}^3\,\text{ster}. \tag{44}$$

Normalizing this to the full Regge-Drell prediction of $0.14\ \mu$b/GeV^3 ster, one arrives at an enhancement factor expressible in terms of the width

$$E_{\varrho'} \approx \frac{0.11}{[\Gamma_{\varrho'}^{\text{tot}}\ \text{in GeV}]^2}, \tag{45}$$

which confirms our earlier expectation.

We want to stress once more: these considerations refer to the p wave part and assume the existence of ϱ' around 1500 MeV. Experimentally we do not know at all what partial waves are involved. In particular, the nature of the shoulder is completely open. All that we wanted to demonstrate is the fact that F.E.S.R. suggests strong p wave suppression at higher dipion masses.

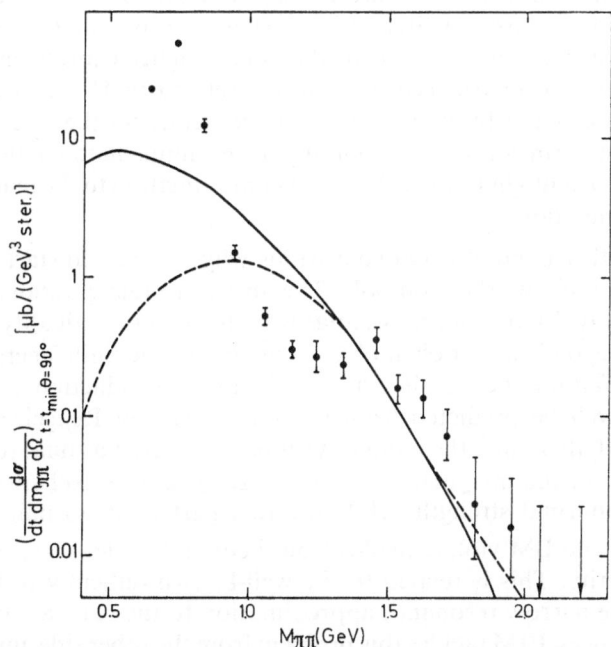

Fig. 18. Plot of $d\sigma/dt_{pp}dm_{\pi\pi}d\Omega$ versus $m_{\pi\pi}$ for $t_{pp}=t_{\min}$, Jackson angle $\theta = 90°$. The solid curve is the Reggeized Drell model. The broken curve is the contribution from the p wave of the imaginary part. The curves are calculated for $E_{\gamma}^{lab} = 15$ GeV on a proton target with the free parameter $v_0 = 0.7$ GeV². The data points are taken from Ref. [38], which is on beryllium and which gives no normalization; we have applied a normalization appropriate for a proton target [5] of $d^2\sigma/dt_{pp}dm_{\pi\pi}d\Omega = 48\,\mu b/\text{GeV}^3$ ster at the ϱ peak for $\theta = 90°$. Figure from Ref. [37]

The smallness of ϱ' production is in no contradiction to the fact that the $1/t_{pp}^2$ behaviour of the proton electromagnetic form factor calls for a strong contribution (ϱ' or non-resonating) of opposite sign to the ϱ contribution [40]. In particular, it does not demand $g_{\gamma\varrho'}/g_{\gamma\varrho}$ (which are defined at $k^2 = m_{\varrho}^2, m_{\varrho'}^2$) to be small. For such conclusion, one would require strong assumptions on mass continuations, on completely unknown quantities as $\sigma_{\varrho'p}^{tot}$ as well as the diffractive transition $\sigma(\varrho'p \rightarrow \varrho p)$.

4. Summary and Conclusions

In an attempt to gain some insight into the diffraction dissociation dynamics by concentrating on the relation between resonant and non-resonant 2π photoproduction, we discussed three models, all of which describe the gross features of the empirical $m_{\pi\pi}$ and t_{pp} distributions below $m_{\pi\pi} = 1$ GeV equally well: the interference model (IM), the final-state interaction model (FIM) and the B_4 model (BM). IM obtains the observed deviation from a Breit-Wigner shaped ϱ by an additive, FIM by a multiplicative modification of the Breit-Wigner expression, while BM is somewhere in between, but closely related to IM insofar as it is well approximated by its π and ϱ pole contents. So the interference model (in the form ϱ pole $+ \pi^{\mp}$ poles) is in no immediate conflict with duality. In fact, it might be regarded as an approximation to the manifestly dual B_4 formulation.

On the other hand, it is not so obvious why one should start from a B_4 form that includes the pion pole. The answer to this question, which concerns the real parts in an essential way, is outside of duality and it goes much beyond the problem of $\pi - \varrho$ duality in the finite energy sum rule sense. Within the FESR framework, the ϱ production rate can reasonably well be predicted from the known size of Reggeized pion exchange and, therefore, the ϱ might well be considered as dual to the π. In other words, the imaginary part of the Reggeized π exchange (with $\alpha'_{\pi} = 1$) is of normal strength, while the real part is abnormally large.

Both IM and BM violate the Watson theorem for $\gamma\, \mathbb{P} \to \pi\pi$, i.e., they violate unitarity. This is related to the well-known difficulty of how to pass from the narrow resonance approximation to the real case of finite width resonances. FIM tackles this problem from the other side, imposing the correct phase from the very beginning. Both ends should meet as soon as there is progress in unitarization of dual amplitudes.

Further progress in the understanding of 2π photoproduction will come from more detailed studies of structures in the $m_{\pi\pi}$ dependence of observable quantities like the t_{pp} slope around the ϱ or by performing experiments with reasonable statistics in the higher dipion mass region, which is still rather unexplored but is very important in the study of non-resonant 2π production (e.g., everybody somehow believes in the Drell mechanism, but nobody has seen multiperipheralism at its best, namely when all subenergies are large). As it stands, the process $\gamma p \to \varrho p$ is just a Gedankenexperiment, which is not known to better accuracy than the models that we have on $\gamma p \to \pi^{+} \pi^{-} p$.

Acknowledgements. It is a pleasure to thank P. H. Frampton, G. Kramer, H. Satz and C. Schmid for many interesting discussions.

References

1. The first authors suggesting this were: Berman, S. M., Drell, S. D.: Phys. Rev. **133**, B 791 (1964).
2. See e.g.: Satz, H.: Rapporteur's Talk at the 1971 Amsterdam International Conference on Elementary Particles. — Satz, H., Schilling, K.: Aspects of Diffraction Dissociation, to appear in the Proceedings of the Colloquium on Multiparticle Dynamics, Helsinki 1971.
3. Some authors even speculate that diffractive dissociation cross-sections diverge logarithmically. See e.g.: Stodolsky, L.: SLAC-Pub 864 (1971). — Cheng, H., Wu, T. T.: DESY Preprint 71/36.
4. See e.g.: Kittel, W.: Review of experimental results on few body reactions, to appear in the Proceedings of the Colloquium on Multiparticle Dynamics, Helsinki 1971 and D.R.O. Morrison, Ref. [6].
5. See e.g.: Wolf, G.: Photoproduction of vector mesons. Invited Talk presented at the Symposium on Meson Photo- and Electroproduction at Low and Intermediate Energies, Bonn 1970, DESY Preprint 70/64.
6. Deck, R. T.: Phys. Rev. Letters **13**, 169 (1964). — Chew, G., Pignotti, A.: Phys. Rev. Letters **20**, 1078 (1968). — Berger, E. L.: Contribution N° 22 to the Amsterdam Conference, 1971. — Morrison, D. R. O.: Rapporteur's Talk at the XV th International Conference on High Energy Physics, Kiev 1970.
7. Joos, P.: Compilation of Photoproduction data above 1.2 GeV. DESY-HERA 70–1.
8. Badier, S., Bouchiat, C.: Phys. Letters **15**, 96 (1965).
9. Kramer, G., Schilling, K.: Z. Physik **191**, 51 (1966). — Freund, P. G. O.: Nuovo Cimento **44**, 411 (1966). — Bucella, F., Colocci, M.: Phys. Letters **24** B, 61 (1967). — Joos, H.: Phys. Letters **24** B, 103 (1967). — Kajantie, K., Trefil, J. S.: Phys. Letters **24** B, 106 (1967).
10. Freund, P. G. O.: Nuovo Cimento **48**, 541 (1967). — Barger, V., Cline, D.: Phys. Rev. Letters **24**, 1313 (1970).
11. Schilling, K., Storim, F.: Nucl. Phys. B **7**, 559 (1968).
12. Hilpert, H. G., et al.: Nucl. Phys. B **23**, 45 (1970). — Eisenberg, Y., et al.: Weizmann Institute Preprint, 1971.
13. McClellan, G., et al.: Cornell Preprint 1971.
14. Hilpert, H. G., et al.: Nucl. Phys. B **21**, 193 (1970).
15. Cohen-Tannoudji, G., Salin, Ph., Morel, A.: Nuovo Cimento **55**, 412 (1968).
16. Schilling, K., Seyboth, P., Wolf, G.: Nucl. Phys. B **15**, 397 (1970).
17. SLAC-Berkeley-Tufts collaboration: Phys. Rev. Letters **23**, 498 (1970); **24**, 955, 960, 1364 (1970).
18. For a more detailed discussion of the topic of s and t channel helicity conservation see e.g.: Satz, H., Schilling, K.: Ref. [2].
19. Gotsman, E., Mannheim, P. D., Maor, U.: Phys. Rev. **186**, 1703 (1969).
20. Mannheim, P. D., Maor, U.: Phys. Rev. D **2**, 2105 (1970).
21. Derado, I., Kronseder, G., Schacht, P., Schlamp, P.: Max-Planck-Institut, München. Preprint 1971.
22. Ross, M., Stodolsky, L.: Phys. Rev. **149**, 1172 (1966).
23. Drell, S. D.: Phys. Rev. Letters **5**, 278 (1960).
24. Pumplin, J.: Phys. Rev. D **2**, 1859 (1970).
25. Dewey, P., Humpert, B.: Cavendish Laboratory Preprint 1971.
26. Kramer, G., Uretsky, J.: Phys. Rev. **181**, 1918 (1969). — Kramer, G., Quinn, H. R.: Nucl. Phys. B **27**, 77 (1971).
27. Pokorski, S., Satz, H.: Nucl. Phys. B **19**, 113 (1970).
28. Satz, H., Schilling, K.: Nuovo Cimento **67** A, 511 (1970).
29. Söding, P.: Phys. Letters **19**, 702 (1966).

30. See e.g.: Schmid, C.: Phenomenology at intermediate energies. Invited Talk at the Conference on the Phenomenology of Particle Physics, CALTECH 1971, CERN Preprint TH. 1341 (1971).
31. Bauer, T.: Phys. Rev. Letters **25**, 485 (1970).
32. It is amusing to note that the normalization

$$\langle E \rangle = \langle \mathrm{Im}\, E \rangle = \int\limits_{m_\varrho^2 - a}^{m_\varrho^2 + a} d\, m_{\pi\pi}^2\, E(m_{\pi\pi}) = 1$$

over the resonance spacing, yielding $\lambda = 1/\pi$, correctly predicts the experimental cross-section size.
33. Dolen, R., Horn, D., Schmid, C.: Phys. Rev. Letters **19**, 402 (1967).
34. Veneziano, G.: Nuovo Cimento **57**, 190 (1968). — Lovelace, C.: Phys. Letters **28**B, 264 (1968). — Shapiro, J.: Phys. Rev. **179**, 1345 (1969).
35. Bartl, A., Iso, C.: DESY Preprint 70/56.
36. Dorren, D., Rittenberg, V., Yaffee, D.: Nucl. Phys. B**30**, 306 (1971).
37. I am reporting here on some recent work of: Frampton, P. H., Schilling, K., Schmid, C.: CERN Preprint TH. 1347 (1971), to be published in Phys. Letters.
38. Bulos, F., *et al.*: Phys. Rev. Letters **26**, 149 (1971).
39. Alvensleben, H., *et al.*: Phys. Rev. Letters **26**, 273 (1971).
40. Frampton, P. H.: Phys. Rev. **186**, 1419 (1969); D**1**, 3141 (1970). — di Vecchia, P., Drago, F.: Nuovo Cimento Letters **1**, 917 (1969).
41. ABBHHM-Collaboration: Phys. Rev. **175**, 1669 (1968).

Dr. K. Schilling
CERN
CH-Geneva, Switzerland

Vector Meson Dominance, Photo- and Electroproduction from Nucleons*

D. SCHILDKNECHT

Contents

1. Basic Assumptions of the Vector Dominance Model

1.1. Generalities on Electromagnetic Interactions

Electromagnetic interactions are described by a coupling between the photon field operator $A_\mu(x)$ and the electromagnetic current operator $j_\mu^{el.}(x)$, containing a leptonic and a hadronic part:

$$j_\mu^{el.}(x) = j_\mu^{lept.}(x) + j_\mu^{hadr.}(x) . \tag{1}$$

The interaction between photons and leptons, as described by QED, is of interest in this lecture in so far only, as this coupling yields a source

---

* Invited Lectures given at the International Summer Institute in Theoretical Physics, Hamburg 1971.

of real or virtual (timelike or spacelike) photons. Thus we simply write in what follows

$$j_\mu \equiv j_\mu^{\text{hadr.}} \,. \tag{2}$$

Let us note some further properties of j_μ. Neglecting weak interactions, the hadronic part of the electromagnetic current is conserved by itself

$$\partial^\mu j_\mu(x) = 0 \,. \tag{3}$$

As regards isospin properties, in analogy to the Gell-Mann Nishijima relation

$$Q = \tfrac{1}{2} Y + I^{(3)} \tag{4}$$

where Q is the charge, Y the hypercharge and $I^{(3)}$ the third component of isospin, we assume a decomposition of j_μ into an isoscalar $(j_\mu^{(Y)})$ and an isovector $(j_\mu^{(I_3)})$ part

$$j_\mu(x) = \tfrac{1}{2} j_\mu^{(Y)} + j_\mu^{(I_3)} \,. \tag{5a}$$

Within $SU(3)$ symmetry, $j_\mu(x)$ transforms as the U spin scalar component of the octet

$$j_\mu(x) = 3^{-1/2} j_\mu^{(8)}(x) + j_\mu^{(3)}(x) \,,$$
$$(\tfrac{1}{2} j_\mu^{(Y)} \equiv 3^{-1/2} j_\mu^{(8)}, \; j_\mu^{(3)} \equiv j_\mu^{(I_3)}) \,. \tag{5b}$$

From (5) we have $|\Delta I \leq 1|$. For a discussion of the experimental situation as regards the possible existence of an $I = 2$ (exotic) electromagnetic current we refer to Donnachie [1].

Describing the electromagnetic interactions of the hadrons in lowest order of the coupling constant $e(e^2/4\pi = \alpha \cong 1/137)$, we have to evaluate the matrix element of $j_\mu(x)$ between hadron states $|A\rangle, |B\rangle$ characterized by particle quantum numbers and momenta

$$T_\mu(\gamma A \to B) \equiv \langle B| j_\mu(0) |A\rangle \,, \tag{6a}$$

which matrix element is determined by the strong interactions solely. Lacking a fundamental theory of hadrons from which we could evaluate (6) by a well founded approximation scheme, in order to proceed further we have to make simple assumptions, which preferably should be such that they may be most directly confronted with experiment. Such assumptions are being made within the framework of vector meson dominance, which will be described in what follows.

1.2. Existence of Vector Meson

From experiments we know for almost ten years [2] now that there exist shortly living strongly interacting particle states, which apart from their mass have the same quantum numbers as the photon, the vector mesons ϱ^0, ω, and ϕ. Some properties [3] of these particles are listed in Table 1.

Having the photon quantum numbers, ϱ^0, ω, ϕ may be produced by direct transition from virtual timelike photons of the correct mass. Indeed, during the last few years ϱ^0, ω, ϕ production has been observed

Table 1

Particle	Main decay	Mass MeV	Width MeV	J^P	C	I
ϱ^0	$\pi\pi \cong 100\%$	765 ± 10	125 ± 20	1^-	-1	1
ω	$\pi^+\pi^-\pi^0 \cong 90\%$ $\pi^0\gamma \cong 10\%$	783.9 ± 0.3	11.4	1^-	-1	0
ϕ	$K^+K^- \cong 46\%$ $\pi^+\pi^-\pi^0 \cong 18\%$ $K_L K_S \cong 35\%$	1019.5 ± 0.6	4.0	1^-	-1	0

[4] in colliding beam experiments at Orsay and Novosibirsk according to

$$e^+ e^- \rightarrow \varrho^0 \rightarrow 2\pi,$$

$$\omega \rightarrow 3\pi,$$

$$\phi \rightarrow K^+ K^-$$

as dominating maxima in the corresponding channels.

It is thus an experimental fact that ϱ^0, ω, ϕ are dominating $e^+ e^-$ annihilation up to energies of approximately 1.1 GeV. In the framework of vector meson dominance this property of dominating vector meson contributions is generalized from the matrix element $\langle 2\pi| j_\mu^{(3)}(0) |0\rangle|_{k^2 \approx -m_\varrho^2}$ to the matrix elements of j_μ between arbitrary hadron states $|A\rangle$, $|B\rangle$ and for arbitrary four momentum transfers $k^2 = (p_B - p_A)^2$ (p_A four momentum of A, p_B four momentum of B) especially for $k^2 = 0$ and for spacelike values $k^2 > 0$: As formulated more precisely in the following sections, the whole k^2 dependence of the matrix elements of the electromagnetic current is assumed to be essentially due to the vector meson poles.

1.3. Current Field Identity

To formulate the basic relations of vector meson dominance [5], let us first write down the matrix element of the hadron reaction $VA \rightarrow B$, where V is a vector meson and A and B are arbitrary hadrons. The amplitude describing this reaction from well known reduction formulae is given by

$$T_\mu(VA \rightarrow B) = \langle B| J_\mu^{(V)}(0) |A \rangle, \tag{6b}$$

where $J_\mu^{(V)}(x)$ is the current to which the vector meson V is coupled, i.e.

$$(\partial^\varrho \partial_\varrho - m_V^2) V_\mu(x) = J_\mu^{(V)}(x) \tag{7}$$

and $V_\mu(x)$ denotes the vector meson field. $J_\mu^{(V)}(x)$ is assumed to be conserved

$$\partial^\mu J_\mu^{(V)}(x) = 0, \tag{8}$$

which is important for the extrapolation in k^2 from the vector meson mass $k^2 = -m_V^2$ to $k^2 \geq 0$ to be discussed later on. Forming matrix elements of (7) with translation invariance we have

$$\langle B| V_\mu(0) |A \rangle = - \langle B| J_\mu^{(V)}(0) |A \rangle / (k^2 + m_V^2). \tag{9}$$

Using now the different parts (5a, b) of the electromagnetic current as interpolating fields of the vector mesons (current field identity) in (9)

$$j_\mu(x) = \sum_{V = \varrho^0, \omega, \phi} (-em_V^2/2\gamma_V) V_\mu(x), \tag{10}$$

we obtain a simple connection between the matrix elements of photon and vector meson induced reactions

$$\begin{aligned} T_\mu(\gamma A \rightarrow B) &= \langle B| j_\mu(0) |A \rangle \\ &= \sum_V (e/2\gamma_V)(m_V^2/(k^2 + m_V^2)) \langle B| J_\mu^{(V)}(0) |A \rangle. \end{aligned} \tag{11}$$

The right hand side has to be evaluated at the value of the four momentum squared $k^2 = (p_B - p_A)^2$ of the real or virtual photon on the left hand

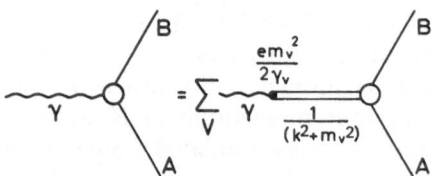

Fig. 1. Graphical representation of Eqs. (11), connecting real or virtual photoprocesses with virtual vector meson induced reactions

side, introducing finite width corrections in the propagator, when using (11) near $k^2 = -m_V^2$. (11) may be represented by the diagram of Fig. 1, i.e. real or virtual photons interact with hadrons via the three vector mesons ϱ^0, ω and ϕ. Relation (11) will be supplemented with certain smoothness assumptions on the variation of the transverse and longitudinal vector meson current matrix elements on the right hand side, which will be specified below.

The coupling constants $\gamma_{\varrho^0}, \gamma_\omega, \gamma_\phi$ are related within $SU(3)$ symmetry by

$$1/\gamma_{\varrho^0} = \sqrt{3}/\gamma_\omega \sin\theta = -\sqrt{3}/\gamma_\phi \cos\theta, \qquad (12)$$

where θ is the $\omega\phi$ mixing angle

$$\begin{aligned}
\phi &= \cos\theta \cdot \phi^8 + \sin\theta \cdot \omega^1, \\
\omega &= -\sin\theta \cdot \phi^8 + \cos\theta \cdot \omega^1.
\end{aligned} \qquad (13)$$

For "ideal mixing", i.e. $\mathrm{tg}\,\theta = 1/\sqrt{2}$, one obtains from (12) the well known relation

$$1/\gamma_\varrho : 1/\gamma_\omega : 1/\gamma_\phi = 3 : 1 : (-\sqrt{2}). \qquad (14)$$

1.4. Smoothness Assumptions of the Vector Meson Dominance Model

Let us formulate now the assumptions on the behaviour of the right hand side of (11), when varying k^2 from the vector meson mass $k^2 = -m_V^2$ to $k^2 = 0$ or to spacelike values $k^2 > 0$:

a) The coupling constants $\gamma_V (V = \varrho^0, \omega, \phi)$ are assumed to be constants independent of k^2 as determined at $k^2 = -m_V^2$ in e.g. e^+e^- annihilation experiments.

b) Transverse amplitudes for $VA \to B$ formed with respect to the s channel helicity frame are assumed to be independent of k^2 for sufficiently high energies

$$\begin{aligned}
s &\equiv -(k + p_A)^2 \gg M_A^2 + m_\varrho^2, \\
s &\gg M_A^2 + k^2,
\end{aligned} \qquad (15a)$$

i.e. with increasing spacelike k^2 in general we expect the domain of validity of the k^2 independence of the model to be shifted to larger and larger values of s, (e.g. Ref. [6]) or in terms of ω

$$\omega \equiv -2P_A k/k^2 \gtrsim 10. \qquad (15b)$$

Let us write down this smoothness assumption somewhat more elaborately. For this purpose we introduce linear polarization states for the vector meson $e_\mu^{(i)}$ fulfilling

$$\begin{aligned}
e_\mu^{(i)} e^{(j)\mu} &= g^{ij}, \qquad i,j = x, y, z \\
k_\mu e^{(i)\mu} &= 0, \qquad (k^2 = -m_V^2)
\end{aligned} \qquad (16)$$

The states $e_\mu^{(x,y)}$ and $e_\mu^{(z)}$ are chosen to be transverse and longitudinal with respect to the c.m.s. of the reaction $VA \to B$ respectively, i.e.

$$\begin{aligned}
e^{(x,y)} &\perp k, \\
e^{(z)} &\parallel k
\end{aligned} \qquad (17)$$

with $e^{(x)}$ in and $e^{(y)}$ perpendicular to the production plane, if $VA \to B$ is a simple two body reaction $VA \to b + c$. The smoothness assumption then states that

$$e_\mu^{(x,y)} T^\mu(k^2, s, \ldots) = e_\mu^{(x,y)} T^\mu(-m_V^2, s, \ldots), \tag{18}$$

or in other words the transverse current matrix elements

$$\langle B| J_\perp^{(V)}(0) |A\rangle \tag{19}$$

are assumed to be independent of k^2 for sufficiently high energies.

c) For the longitudinal amplitudes [7–9] the smoothness assumption cannot be made in the simple form stated above for the transverse ones, as from current conservation

$$k_\mu T^\mu(VA \to B) = 0 \tag{20}$$

the longitudinal amplitude has to go to zero for $k^2 \to 0$. Indeed from (20) and from the expression for the longitudinal polarization vector

$$e_\mu^{(z)} = (1/\sqrt{-k^2})(|\boldsymbol{k}|, k^0 \boldsymbol{k}/|\boldsymbol{k}|), \quad k^2 < 0$$
$$e_\mu^{(z)} e^{(z)\mu} = 1 \tag{21}$$

we obtain

$$e_\mu^{(z)} T^\mu = \sqrt{-k^2} \langle B| J_\parallel^{(V)}(0) |A\rangle / k^0 \tag{22}$$

with

$$\langle B| J_\parallel^{(V)}(0) |A\rangle \equiv T^\parallel = (\boldsymbol{k}/|\boldsymbol{k}|) \langle B| \boldsymbol{J}^{(V)}(0) |A\rangle. \tag{23}$$

It is thus tempting to require the right hand side to be independent of k^2 in (22) only after having taken out the factor $\sqrt{-k^2}$. More elaborately, let us introduce the timelike polarization vector

$$e_\mu^{(0)} = -e_{(0)\mu} = (1/\sqrt{k^2})(|\boldsymbol{k}|, k^0/|\boldsymbol{k}| \boldsymbol{k}),$$
$$k^2 > 0, \tag{24}$$

and let us formulate the smoothness requirement for the longitudinal amplitude as [8][1]

$$(e_{(0)\mu} T^\mu/\sqrt{k^2})_{k^2 > 0} = (e_\mu^{(z)} T^\mu/\sqrt{-k^2})_{k^2 = -m_V^2} \tag{25}$$

---

[1] There are slight differences in the mass extrapolation procedure between different authors. The procedure [8] described here, has the virtue of being invariant against Lorentz transformations in the direction of $\boldsymbol{k}$. Other procedures [7, 9, 56] require the not Lorentz invariant quantity $\langle B| J_\parallel^{(V)} |A\rangle$ to be mass independent with respect to e.g. the laboratory [7] or the c.m.s. [56] and take into account the variation of $\sqrt{-k^2}/k^0$ with k^2. Different procedures are equivalent for $s \gg m_V^2, k^2$. Compare also the brief discussion on the mass extrapolation procedure in the Born term model chapter 3.2. and Ref. [60]. The question of why to chose $e_{\mu(0)} T^\mu$ to be smooth and not $e_\mu^{(0)} T^\mu$ is discussed in Ref. [67].

which equation with (22) means that the Lorentz invariant parallel current component matrix element is assumed to be smooth:

$$(\langle B| J_{\|}^{(V)}(0) |A\rangle /k^0)_{k^2} = (\langle B| J_{\|}^{(V)}(0) |A\rangle /k^0)_{-m_V^2}. \tag{26}$$

The smoothness assumptions allow us to directly connect amplitudes for reactions induced by real or virtual photons with reactions induced by on mass shell vector mesons $e_\mu T^\mu(V A \to B)$:

$$(e_\mu^{(x,y)} T^\mu(\gamma A \to B))_{k^2 \geq 0} = \sum_V \frac{\sqrt{\alpha \pi}}{\gamma_V} \frac{m_V^2}{(k^2 + m_V^2)} e_\mu^{(x,y)} T^\mu(V A \to B),$$

$$(e_{(0)\mu} T^\mu(\gamma A \to B))_{k^2 \geq 0} = \sum_V \frac{\sqrt{\alpha \pi}}{\gamma_V} \frac{m_V^2}{(k^2 + m_V^2)} \frac{\sqrt{k^2}}{m_V} e_\mu^{(z)} T^\mu(V A \to B). \tag{27}$$

Eqs. (27) are the fundamental relations of vector meson dominance, thus relating cross sections for photon reactions, e.g.

$$\gamma N \to \pi N$$
$$\gamma N \to \varrho^0 N \tag{28}$$

to cross sections of reactions induced by vector mesons

$$(\varrho^0, \omega, \phi) N \to \pi N,$$
$$\varrho^0 N \to \varrho^0 N. \tag{29}$$

With additional more or less well verified assumptions, (29) may be related to reactions actually measurable in the laboratory as

$$\pi N \to (\varrho^0, \omega, \phi) N,$$
$$\pi N \to \pi N, \tag{30}$$

thus predicting photon reactions from vector meson induced hadron reactions (vector meson photon analogy).

Let us add a remark at this point on the choice of the s channel helicity frame when formulating the smoothness assumptions of the vector meson dominance model. A priori there is no specific reason [10] to prefer this frame to any other one. Because of the spin rotation of massive particles under Lorentz transformations the predictions for e.g. real photoproduction reactions will in general be different, when chosing a different frame to make the smoothness assumptions of the VDM. This ambiguity is only clarified by looking at specific dynamical models. The choice of the s-channel helicity frame may thus be motivated e.g. for single pion production by studying the electric Born term model [10–12]. Maximal smoothness in this model is obtained at sufficiently high energies with respect to the s channel helicity frame.

Instead of starting by making the smoothness assumptions in terms of transverse and longitudinal amplitudes, one may, of course, decompose

$T_\mu(VA \to B)$ in terms of invariant amplitudes, chosen to be free from kinematic singularities and postulate [10] the smoothness assumption for these invariant amplitudes. I shall come back to this point in Chapter 4, as it leads to additional interesting consequences [13, 14]. At this point, I only wish to mention that for the specific process $VN \to \pi N$ smoothness for the Ball invariant amplitudes [16] leads to the result postulated above, namely smoothness with respect to the s channel helicity frame [15].

1.5. Current Field Identity in Lagrangian Models, Dispersion Relations and Historical Remarks

At this point let us add a few remarks on the historical development of vector meson dominance and on different starting points for the current field identity (10) and the matrix element relation (11).

Whereas *we* could start from the experimental facts of the existence of vector mesons, and their production in $e^+ e^-$ annihilation, the historical approach to vector meson dominance has been much more speculative and difficult.

The existence of vector mesons was first predicted by Nambu [17] in 1957 and Frazer and Fulco [18] in 1959, soon after the first measurements of the nucleon formfactors had become available. Sakurai [19], in 1960, postulated vector mesons on theoretical grounds, when attempting to formulate a theory of strong interactions in close analogy to QED based on the concepts of conserved currents and vector mesons universally coupled to them. Vector mesons were discovered experimentally [2] in the years 1961–1963.

The standpoint of introducing the CFI as a definition of interpolating fields of the vector mesons within the framework of General Field Theory has been emphasized by Joos [20] in 1967. Kroll, Lee and Zumino [21] in 1967 showed that it is possible to construct a Lagrangian field theory, fully satisfying gauge invariance in which the CFI is obtained as a consequence of the equations of motion. Instead of introducing the CFI, the matrix element relation (11) may also be obtained by postulating dispersion relations in the mass variable and saturating these relations with the vector meson poles. This approach to vector meson dominance has been pioneered by Gell-Mann and Zachariasen [22] in 1961.

1.6. General Remarks on Tests of Vector Meson Dominance

Qualitatively, from the vector meson dominance relations (27) we expect photons to behave hadronlike, i.e. the same general features as a function of the kinematic variables are expected for photoproduction and vector meson induced reactions.

More quantitative tests of the ansatz (27) are obtained by comparing the relative normalization of photon and vector meson induced reactions using the $e^+ e^-$ annihilation values of the coupling constants $\gamma_{\varrho^0}, \gamma_\omega, \gamma_\phi$. For real photons (27) becomes

$$e_\mu^{(x,y)} T^\mu(\gamma A \to B)_{k^2 = 0} = \sum_V (\sqrt{\alpha\pi}/\gamma_V) \, e_\mu^{(x,y)} T^\mu((VA \to B)), \qquad (31)$$

and the relations between cross sections, as they are obtained from this formula, have been compared for numerous experiments during the last few years.

Measurements of the k^2 dependence as predicted in (27) – apart from the well known cases of the nucleon elastic form factors and the total electroproduction cross section – have just given the first results e.g. for $\gamma_{\text{virt.}} N \to \pi N, \pi \Delta, \varrho^0$ in the high energy region during the last few months. Whereas (31) essentially tests whether the residues of the vector meson poles as obtained from measurements with real photons have the correct values as predicted from vector meson reactions, electroproduction experiments directly allow to test the predicted pole behaviour in k^2 over a large range of $k^2 > 0$ in an extremely sensitive way.

Of special interest for the concepts of vector meson dominance is the exploration of $e^+ e^-$ annihilation at energies beyond the masses of ϱ^0, ω, ϕ which has only started rather recently. Important questions, like existence of higher mass vector mesons, and the production mechanism for multibody final states, whether e.g. production proceeds via the tails of ϱ^0, ω, ϕ, can thus be experimentally investigated.

2. The Coupling Constants $\gamma_\varrho, \gamma_\omega, \gamma_\phi$ Brief Remarks on $e^+ e^-$ Annihilation beyond ϱ^0, ω, ϕ

The most recent values for the ϱ^0 width and the coupling constant γ_ϱ have been obtained in an Orsay experiment [23], which was designed to especially measure the $\varrho^0 \omega$ electromagnetic mixing effect [24]. The values obtained [23] are given in Table 2 in comparison with previous

Table 2

	Orsay I	Orsay II	Novosibirski S-wave B.W. formula
m_{ϱ^0}[MeV]	776.0 $\pm$ 6	780.2 $\pm$ 5.9	754 $\pm$ 9
Γ_{ϱ^0}[MeV]	127.3 $\pm$ 12.5	152.8 $\pm$ 15.1	105 $\pm$ 20
$B_{\varrho^0 \to e^+ e^-} \cdot 10^{-5}$	6.2 $\pm$ 0.6	4.0 $\pm$ 0.36	5 $\pm$ 1
$\Gamma_{\varrho^0 \to e^+ e^-}$ [keV]	7.9 $\pm$ 7	6.11 $\pm$ 0.53	5.2 $\pm$ 0.5
$\gamma_{\varrho^0}^2/4\pi$	0.51 $\pm$ 0.05	0.64 $\pm$ 0.06	

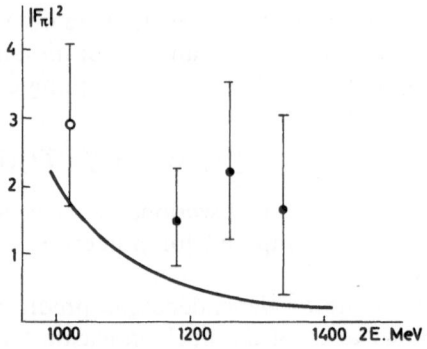

Fig. 2. The square of the pion from factor as obtained at Orsay (◊) and Novosibirsk (♦). The curve shows the ϱ^0 dominance prediction (Fig. from Sidorov, Ref. [4])

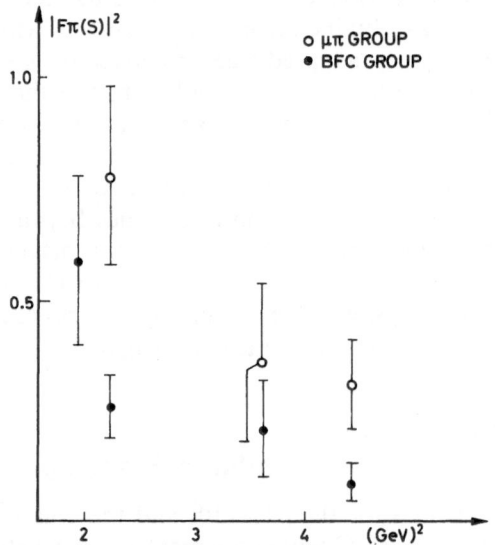

Fig. 3. The square of the pion formfactor as obtained at Frascati. (Fig. from Ref. [27])

results. The Orsay ϱ^0 data have been fitted with the Gounaris Sakurai formula [25].

The results for the ω and ϕ photon coupling are

$$\gamma_\omega^2/4\pi = 4.6 \pm 0.5 \ [23]$$
$$\gamma_\phi^2/4\pi = 2.9 \pm 0.2 \ [26] \tag{32}$$

yielding (with $\gamma_\varrho^2/4\pi = 0.64 \pm 0.06$)

$$\gamma_\varrho^{-2} : \gamma_\omega^{-2} : \gamma_\phi^{-2} = 9 : (1.25 \pm 0.25) : (2 \pm 0.4) \tag{33}$$

approximate agreement with $SU(3)$ with ideal $\omega\phi$ mixing.

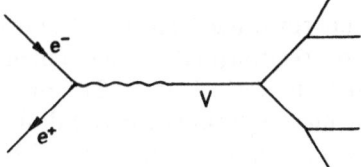

Fig. 4. Mechanism for production of a many particle final state via vector meson dominance

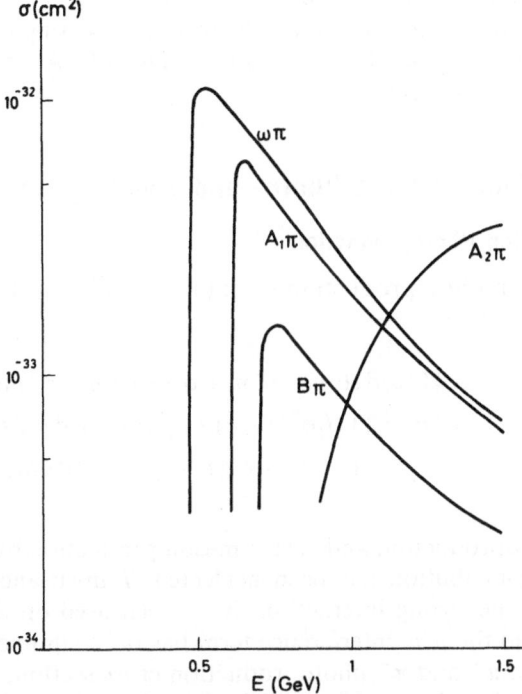

Fig. 5. Vector meson dominance predictions from Ref. [28] for a few dominating channels in e^+e^- annihilation as a function of the electron energy E in the e^+e^- c.m.s.

In the energy range beyond ϱ^0, ω, ϕ, one of the questions of special interest for vector meson dominance is the existance of higher mass vector mesons, in particular of the ϱ' meson, which, if it exists, should of course be included in our basic Eqs. (10) and (11). Because of limited statistics and no clear separation of π and K mesons, as stressed by Bernardini [27] for the Frascati data, experiments so far do not give a reliable answer as regards the existence of a ϱ'.[1] Figs. 2 and 3 show Novosibirsk [4] and Frascati [27] results on $e^+e^- \to \pi^+\pi^-$.

---

[1] *Note added in proof:* More recently, evidence for a possible ϱ' resonance at $m \approx 1.6\,\text{GeV}$, $\Gamma \approx 0.35\,\text{GeV}$ has been reported in $e^+e^- \to \pi^+\pi^+\pi^-\pi^-$ (G. Barbarino *et al.* Lett. Nuovo Cim., A. Bramon and M. Greco, Lett. Nuovo Cim. to be published).

First results in the energy range from 1.4–2.4 GeV $e^+ e^-$ c.m.s. energy indicate [27] a large cross section (of the order of magnitude of 10^{-32} cm^2) for multihadron production, as $e^+ e^- \to 2\pi^+ 2\pi^-$ or $e^+ e^- \to 2\pi^+ \pi^-$ + neutrals. Assuming vector meson dominance and dominance of quasi-two-body channels as shown in Fig. 4, Kramer *et al.* [28] and Layssac *et al.* [29] obtain estimates [30] for cross sections of the order of magnitude measured at Frascati. The predictions for the cross sections for a few presumably dominating channels are shown on Fig. 5 [28]. Let us wait, how more refined experimental results from the existing ring at Frascati and from the new generation of storage rings at SLAC and DESY will compare with these predictions!

3. Single Pion and $\pi\varDelta$ Photo- and Electroproduction

3.1. Single Pion Photoproduction

For single pion photoproduction from (31) one obtains the well known relation [31–33]

$$
\begin{aligned}
\tfrac{1}{2}\,d\sigma/dt(\pi^+ + \pi^-) &\equiv \tfrac{1}{2}(d\sigma/dt(\gamma p \to \pi^+ n) + d\sigma/dt(\gamma n \to \pi^- p)) \\
&= (\alpha\pi/\gamma_\varrho^2)\,[\tfrac{1}{2}(\varrho_{xx}^{\varrho^0}(1-\xi) + \varrho_{yy}^{\varrho^0}(1+\xi))\,d\sigma/dt(\pi^- p \to \varrho^0 n) \\
&\quad + \tfrac{1}{2}(\varrho_{xx}^{\omega}(1-\xi) + \varrho_{yy}^{\omega}(1+\xi))\,\gamma_\varrho^2\gamma_\omega^{-2}\,d\sigma/dt(\pi^- p \to \omega n)]
\end{aligned}
$$

$$(34)$$

between photoproduction and vector meson production by pions, where the small ϕ contribution has been neglected. T invariance and isospin invariance of the strong interactions have been used on the right hand side in (34) and the $\varrho^0\omega$ interference term has been eliminated by taking the sum of the π^+ and π^- photoproduction cross sections from protons and neutrons. ξ with $-1 \leqq \xi \leqq +1$ denotes the degree of linear polarization of the incoming photon, $\xi = +1$ and $\xi = -1$ corresponding to complete linear polarization normal to and in the production plane respectively. The density matrix elements of the outgoing ϱ^0 and ω with respect to the linear base vectors (16), (17) have been denoted by $\varrho_{xx}, \varrho_{yy}$ and are related to the usual helicity state density matrix elements by

$$
\begin{aligned}
\varrho_{xx} &= \varrho_{11} - \varrho_{1-1} \\
\varrho_{yy} &= \varrho_{11} + \varrho_{1-1}
\end{aligned}
$$

$$(35)$$

and furthermore

$$
\begin{aligned}
\varrho_{zz} &\equiv \varrho_{00} \\
\varrho_{xz} &= -\sqrt{2}\,\varrho_{10}.
\end{aligned}
$$

$$(36)$$

For the frequently used asymmetry

$$A(\pi^+ + \pi^-) \equiv \frac{d\sigma_\perp/dt(\pi^+ + \pi^-) - d\sigma_\parallel/dt(\pi^+ + \pi^-)}{d\sigma_\perp/dt(\pi^+ + \pi^-) + d\sigma_\parallel/dt(\pi^+ + \pi^-)} \tag{37}$$

where $d\sigma_\perp$ and $d\sigma_\parallel$ denote the production cross sections for photon polarization normal to and in the production plane respectively, from (34) we have in good approximation

$$A(\pi^+ + \pi^-) \cong \varrho^{\varrho^0}_{1-1}/\varrho^{\varrho^0}_{11} . \tag{38}$$

Relations (34) and (38) have been tested quite extensively [34]. Discrepancies from predictions [32] found [35] for the asymmetry A led to the conjecture [36] that smoothness under mass extrapolation should be required with respect to the Donohue Högaasen frame of reference [37] and not with respect to the s channel helicity frame. Investigations within dynamical models, however, suggest [38] that the s channel helicity frame is the correct one as postulated in chapter 1. For a summary of earlier tests of (34), (38) we refer to previous reviews (e.g. Refs. [39, 40]) and on Fig. 6 only show the most recent result [41] based on new data [42] on $\pi^- p \to \varrho^0 n$ at 15 GeV/c. In this experiment [42] the very forward small $-t$ region was resolved for the first time for $\pi^- p \to \varrho^0 n$, the region in which photoproduction shows the characteristic spike structure [43]. For photoproduction this forward structure for $|t| \lesssim m_\pi^2$ is quite well described [44] by the electric Born term model. From vector meson dominance we expect the same dynamics to also hold for $\pi N \to \varrho^0 N$ and thus expect [11, 12] the same forward structure to show up for transverse ϱ^0 production with respect to the s channel helicity frame, where maximal smoothness holds [10–12] under mass extrapolation at high energies. Fig. 6 shows that the gross features of the photon data [35, 45], in particular the forward spike in the parallel cross section, are quite well reproduced by the hadron data (s channel helicity frame). $\gamma_\varrho^2/4\pi = 0.5$ has been used in Fig. 6, which value is somewhat lower than the most recent $e^+ e^-$ annihilation determination quoted in Chapter 2. The photon data on Fig. 6 are combinations of 16 GeV/c unpolarized cross sections [45] and asymmetries measured [35] at 3.4 GeV/c.[2] Best quantitative agreement is found for parallel photon polarization, whereas discrepancies are present for the normal polarization, thus leading to discrepancies in the asymmetry. These are less severe than the ones observed earlier (see e.g. Refs. [39, 40]), as A computed from (38) as seen on Fig. 6 turns out to be positive throughout. The sensitivity of A against small deviations of $f_{\varrho NN}/f_{\varrho\pi\pi}$ from the univer-

---

[2] The π^+ asymmetry at 11 GeV/c [118] is consistent with the one at 3.4 GeV/c. π^- asymmetry measurements at higher energy would be most important to allow tests of (38) independent of assumptions on the energy behaviour.

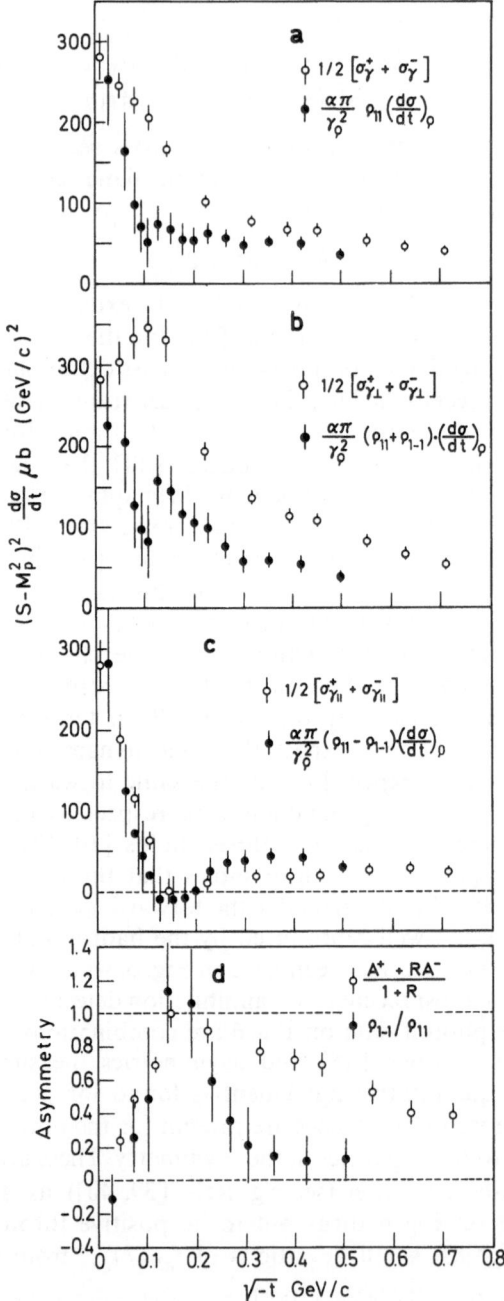

Fig. 6. Comparison [41] of the sum of π^+ and π^- photoproduction cross sections [45, 35] with data on $\pi^- p \rightarrow \varrho^0 n$ [42]. a Unpolarized photoproduction. b Photons polarized normally to production plane. c Photon polarization parallel to production plane. d Asymmetry A

sality [19] ratio $f_{\varrho NN}/f_{\varrho \pi \pi} = 1$ has been stressed in Ref. 46. In connection with the data, it should be noted that the estimated [42] additional overall systematic uncertainty for the normalization of the $\pi^- p \to \varrho^0 n$ data is $\pm 25\%$. Also, the uncertainties arising from the delicate problem of subtracting the $\pi^+ \pi^-$ S-wave production, which is of the same order of magnitude as the transverse cross section, are not included in Fig. 6. Further high statistics results on $\pi^- p \to \varrho^0 n$ are still of great interest in connection with Eq. (34), and are expected to come out in the near future from the CERN Munich collaboration [47].

3.2. Single Pion Electroproduction

Only very recently first data on single pion electroproduction at energies beyond the resonance region were presented by groups from CEA [48], DESY [49] and NINA [50]. Such measurements in particular allow to test the extrapolation procedure for the longitudinal amplitude as given by (27).

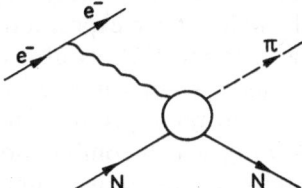

Fig. 7. Single particle (π) electroproduction via one photon exchange

Let us first briefly review the relevant kinematical formulae. For one photon exchange (Fig. 7), the single particle electroproduction cross section (e.g. $eN \to eN\pi, e\Delta\pi, eN\varrho$) is given [51] by

$$d\sigma/dW^2 dk^2 dt\, d\phi = \tfrac{1}{4}\alpha(2\pi)^{-2}(W^2-M^2)\,E^{-2}M^{-2}k^{-2}(1-\varepsilon)^{-1}$$
$$\cdot [\tfrac{1}{2}(\sigma^{xx}+\sigma^{yy})+\varepsilon\sigma^{00}-\varepsilon\cos 2\phi\cdot\tfrac{1}{2}(\sigma^{yy}-\sigma^{xx})$$
$$+\sqrt{2\varepsilon(1+\varepsilon)}\cos\phi\,\sigma^{x0}],$$
$$\varepsilon = [1+2(1+(E-E')^2 k^{-2}\tan^2\tfrac{1}{2}\vartheta]^{-1}, \tag{39}$$

where the different parts of the cross section $\sigma^{ik}(W^2, t, k^2)$ are functions of the square of the invariant mass W of the final hadron state (e.g. $\pi N, \pi\Delta$), of the momentum transfer $t = -(p_\pi - k)^2$ or $t = -(p_\varrho - k)^2$ respectively, and of the virtual photon four momentum squared k^2. ε, as usual, denotes the polarization parameter of the virtual photon, E, E' denote the initial and final electron energies, ϕ the angle between lepton and hadron planes, and ϑ the electron scattering angle in the laboratory. The indices $x, y, 0$ refer to the transverse and longitudinal polarization vectors (16), (17), (24), $\frac{1}{2}(\sigma^{xx}+\sigma^{yy})$ thus denoting the production cross section with unpolarized virtual photons, σ^{00} the production cross section with longitudinally polarized virtual photons etc., the conventional notation being

$$
\begin{aligned}
\sigma_u &\equiv \tfrac{1}{2}(\sigma^{xx}+\sigma^{yy}) \\
\sigma_L &\equiv \sigma^{00} \\
\sigma_T &\equiv \tfrac{1}{2}(\sigma^{xx}-\sigma^{yy}) \\
\sigma_I &\equiv \sigma^{x0}.
\end{aligned}
\tag{40}
$$

Specializing (27), quite analogously to (34) we obtain for $\sigma_L(\gamma_{\text{virt.}} p \to \pi^+ n)$ the prediction [9, 52–54]

$$
\sigma_L = \frac{k^2}{m_\varrho^2} \frac{m_\varrho^4}{(k^2+m_\varrho^2)^2} \frac{\alpha\pi}{\gamma_\varrho^2} \varrho_{00}^0 \frac{d\sigma}{dt} (\pi^- p \to \varrho^0 n) + (\varrho^0, \omega) \text{ interfer.} \tag{41}
$$

where the ω and ϕ contributions have been neglected. From (41), for sufficiently large k^2 and small $-t$, we expect a dominating longitudinal contribution to the cross Section (39), as also observed in $\pi^- p \to \varrho^0 n$. This is actually the case, as $\sigma_u + \varepsilon\sigma_L$ e.g. at $k^2 = m_\varrho^2$ turns out [48–50] to be larger than observed in photoproduction. The t behaviour [49] of the sum $\sigma_u + \varepsilon\sigma_L (\varepsilon \cong 0.75)$, which combination of cross sections has not yet been experimentally separated, is compared with vector meson dominance predictions [54] on Figs. 8a $(k^2 = -0.26 \text{ GeV}^2/c^2)$ and 8b $(k^2 = -0.75 \text{ GeV}^2/c^2)$ at the hadron c.m.s. energy $W^2 = 4.8 \text{ GeV}^2$. (See also Refs. [55, 56] for closely related work). The hadronic cross section entering (41) has been scaled from data [57] at different energies using an $(s - M^2)^{-2}$ behaviour $(\gamma_\varrho^2/4\pi = 0.5)$. The $\varrho^0\omega$ interference term has been estimated [55] by assuming maximal $\varrho^0\omega$ interference and was found to contribute less than $\sim 10\%$ in the t range considered. The small contribution σ_u to the cross Section (39) has been calculated [54] directly from photoproduction, as follows from (27) and (31):

$$
\sigma_u((W^2, t, k^2) = m_\varrho^4(k^2+m_\varrho^2)^{-2} d\sigma/dt(\gamma p \to \pi^+ n). \tag{42}
$$

Photoproduction data from Ref. [58] have been used in (42). The overall features of the data for $\sigma_u + \varepsilon\sigma_L$, in particular the steep t-behaviour, are quite well reproduced by the vector meson dominance predictions. In the quantitative comparison some discrepancies between electroproduction data and predictions are seen, especially for the smaller value of $k^2 = 0.26 \text{ GeV}^2/c^2$ for small $|t|$. Let us remark, however, that we are applying vector meson dominance at rather low energies, where even simple kinematic effects due to the ϱ^0 mass are in general not negligible.

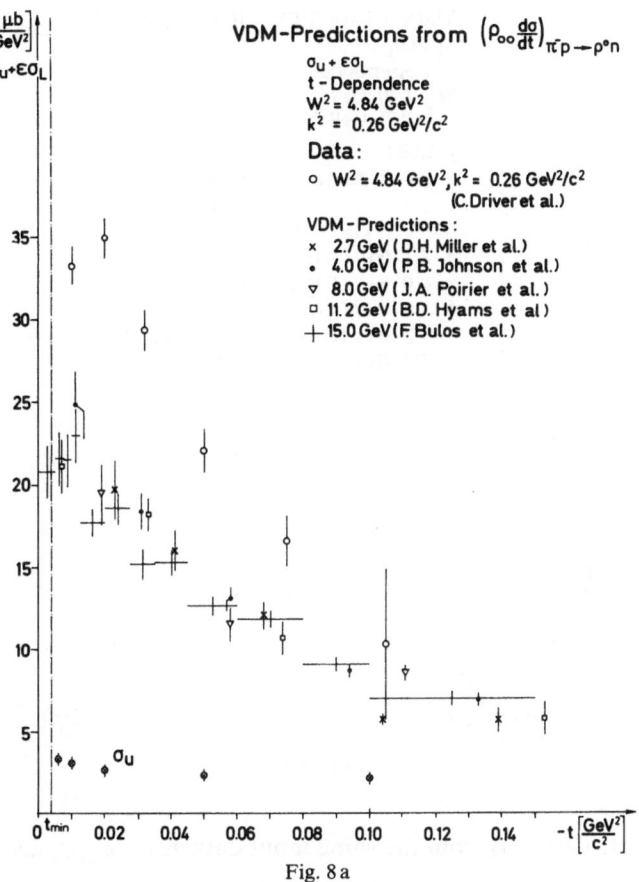

Fig. 8a

Fig. 8. Comparison of electroproduction results [49] for $\sigma_u + \varepsilon\sigma_L$ for $\gamma_{virt}p \to \pi^+ n$ with vector meson dominance predictions from data [57] on $\pi^- p \to \varrho^0 n$ (Fig. from Ref. [54]). a for $k^2 = 0.26$ GeV$^2/c^2$, b for $k^2 = 0.75$ GeV$^2/c^2$

As regards the experimental data, it should be remarked that the data from Ref. [49] lie somewhat higher for small $|t|$ and small k^2 than the data from Ref. [48], the difference being smaller, however, than the difference between data and predictions seen on Fig. 8a.

The k^2 behaviour of σ_T according to

$$\sigma_T = m_\varrho^4 (k^2 + m_\varrho^2)^{-2} (d\sigma_\parallel/dt - d\sigma_\perp/dt)_{\gamma p \to \pi^+ n} \tag{43}$$

as predicted from photoproduction [58] is compared [55] with data [49] on Fig. 9. Assuming that σ_u behaves as predicted in (42), measurements [49] are thus consistent (within large errors) with no significant change in the asymmetry A from (37) when going to spacelike values of k^2.

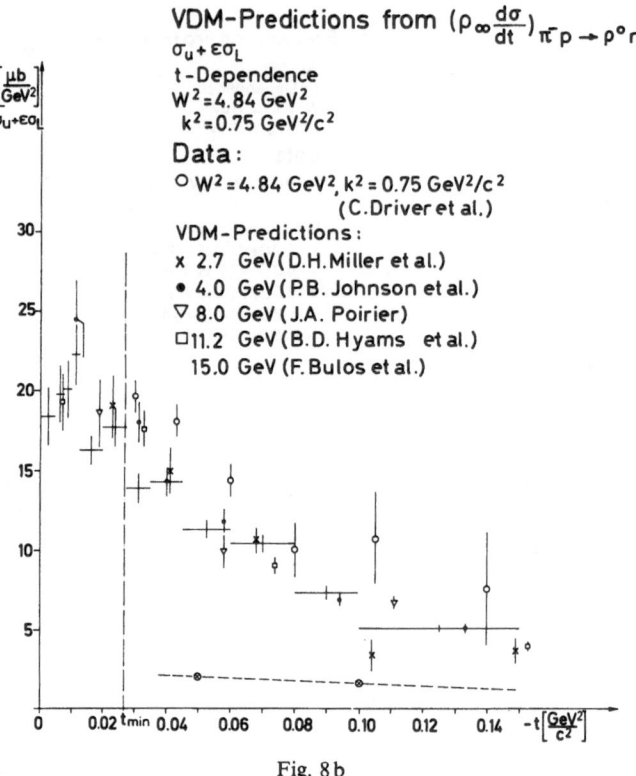

Fig. 8 b

Predictions for σ_I from the same input data as for σ_L, σ_u according to

$$\sigma_I(\gamma_{\text{virt}}p\to\pi^+n)=\left|\sqrt{\frac{k^2}{m_\varrho^2}}\;\frac{m_\varrho^2}{(k^2+m_\varrho^2)^2}\;\frac{\text{Re}\varrho_{xz}^{\varrho^0}}{\frac{1}{2}(\varrho_{xx}^{\varrho^0}+\varrho_{yy}^{\varrho^0})}\;\frac{1}{2}\left(\frac{d\sigma}{dt}(\gamma p\to\pi^+n)\right.\right.$$

$$\left.+\frac{d\sigma}{dt}(\gamma n\to\pi^-p)\right)$$

$$+\sqrt{\frac{k^2}{m_\varrho^2}}\;\frac{m_\varrho^4}{(k^2+m_\varrho^2)^2}\;\frac{\alpha\pi}{\gamma_\varrho\gamma_\omega}\sqrt{\frac{d\sigma}{dt}(\pi^-p\to\varrho^0n)\;\frac{d\sigma}{dt}(\pi^-p\to\omega n)}$$

$$\cdot\left(\sqrt{\varrho_{zz}^{\varrho^0}\varrho_{xx}^{\omega}}+\sqrt{\varrho_{xx}^{\varrho^0}\varrho_{zz}^{\omega}}\right)$$

(44)

are compared with experiment [49] on Fig. 10 [55]. The t behaviour predicted does not show the change of the zero in σ_I with k^2 as observed in electroproduction. This seems to be a kinematical effect, which is explainable [59] within vector meson dominance, when starting from smoothness of invariant amplitudes, treating the kinematical coefficients exactly. Further details will be given in chapter 4.

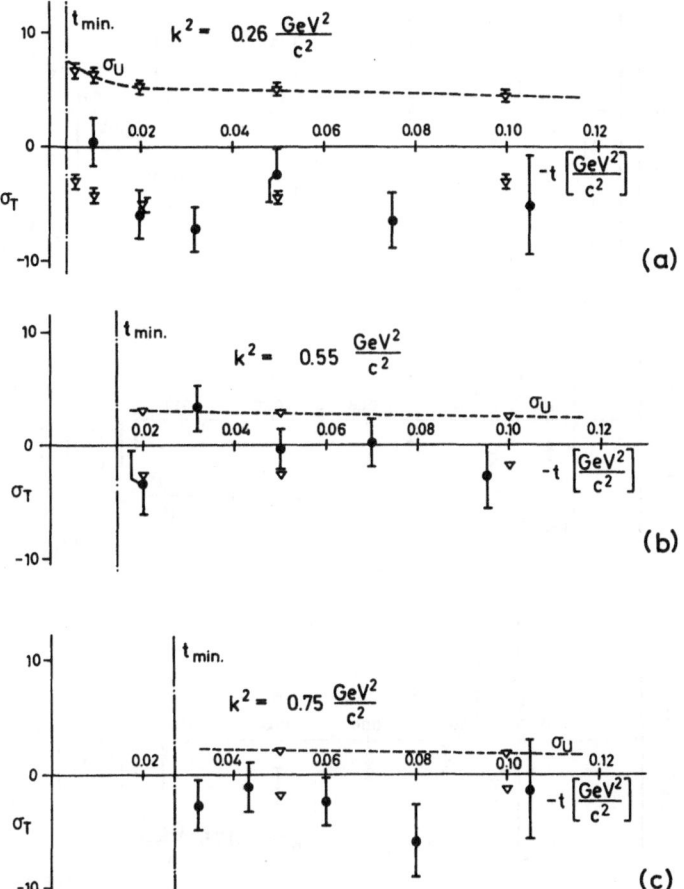

Fig. 9. Comparison [55] of predictions for σ_T for $\gamma_{\text{virt}} p \to \pi^+ n$ according to (43) with electroproduction results for σ_T from Ref. [49]. (VDM predictions: $\triangledown$, data: ϕ). a $k^2 = 0.26 \text{ GeV}^2/c^2$, b $k^2 = 0.55 \text{ GeV}^2/c^2$, c $k^2 = 0.75 \text{ GeV}^2/c^2$

Let us add some brief remarks on calculations for charged pion electroproduction within dynamical models. The electric Born term model (pion exchange plus nucleon pole term with a γ_μ type $\gamma N \bar{N}$ coupling) has recently been used [60] to study the mass extrapolation procedure for the longitudinal amplitudes, i.e. the validity of (25), (26) in comparison with slightly different procedures. The procedure (25), (26) turns out to give the most reasonable approximation of the mass extrapolation as calculated in the Born term model. – Supplementing the Born terms with resonance contributions within the framework of fixed t dispersion relations, several authors [61–63] have extracted

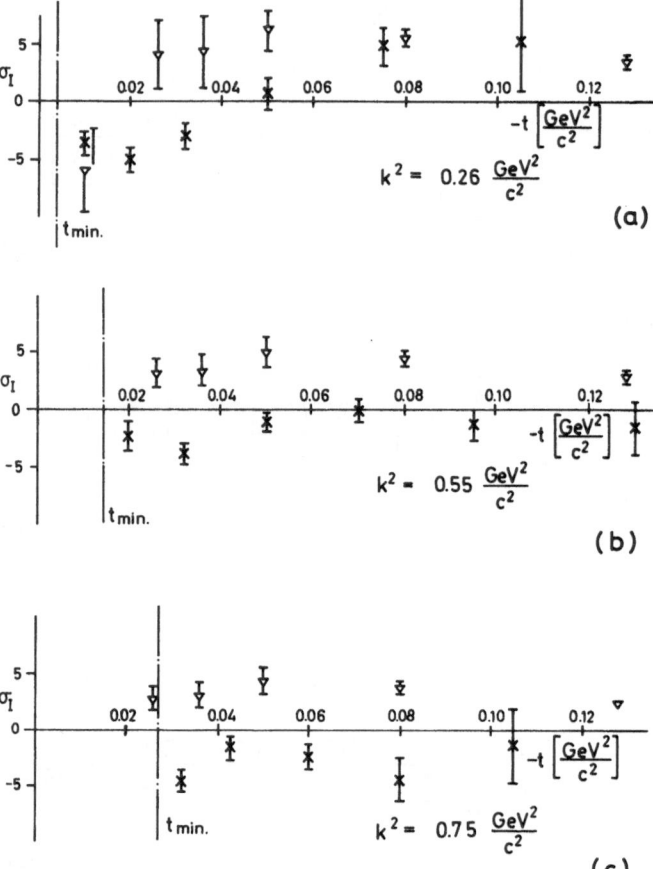

Fig. 10. Comparision [55] of predictions for σ_I for $\gamma_{virt} p \to \pi^+ n$ according to (44) with electroproduction results [49] (VDM predictions: $\vec{\gamma}$, data: ϕ). a $k^2 = 0.26$ GeV$^2/c^2$, b $k^2 = 0.55$ GeV$^2/c^2$, c $k^2 = 0.75$ GeV$^2/c^2$

the pion formfactor from the π^+ electroproduction data. The main contribution to the cross section comes from the Born terms and the $\Delta(1236)$, the effect of higher resonances, which has been investigated in Ref. [63], being quite small. If ϱ^0 dominance is assumed for the pion form factor, the results obtained for $\sigma_u + \varepsilon\sigma_L$ agree with the vector meson dominance predictions shown on Figs. 8a, b. The discrepancy on Fig. 8a for small t also persists in the dispersion theory analysis. The results [62] for the pion form factor are shown on Fig. 11.

Further experiments at higher energies and larger k^2 and a separation of σ_u and σ_L would of course be most important for refined tests of

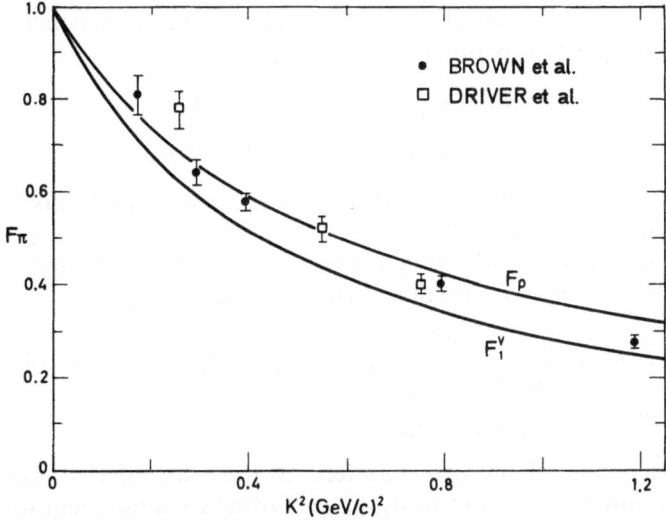

Fig. 11. The pion formfactor as obtained [62] from π^+ electroproduction data [48, 49] above the resonance region via dispersion relations analysis. F_ϱ is the ϱ dominance prediction and F_1^v is the isovector Dirac nucleon formfactor

vector meson dominance predictions for the different parts of the cross section and for a determination of the pion form factor in the spacelike region.

3.3. $\pi \Delta$ (1236) Electroproduction

Vector meson dominance predictions for $\pi \Delta$ (1236) photoproduction [64]

$$\gamma p \to \pi^- \Delta^{++} \tag{45}$$

from

$$\pi^+ p \to \varrho^0 \Delta^{++} \tag{46}$$

have shown [65] quite large discrepancies which may be due, however, to the assumption of line reversal invariance, which has to be made in addition to vector meson dominance in order to predict (45) from the data of reaction (46). A recent calculation [62] within the electric Born term model [66] shows that crossing from the s channel $\pi N \to \varrho \Delta$ to the u channel $\varrho N \to \pi \Delta$ indeed leads to changes of approximately a factor of 2 in the transverse cross sections. No additional crossing assumptions enter, of course, if vector meson dominance is tested by measuring the k^2 dependence of the transverse parts in the $\pi \Delta$ electroproduction cross section, thus checking whether the k^2 dependence due to the ϱ^0 propagator is obtained:

$$\sigma_u = m_\varrho^4 (k^2 + m_\varrho^2)^{-2} \mathrm{d}\sigma/\mathrm{d}t (\gamma N \to \pi \varDelta) \tag{47}$$

$$\sigma_T = m_\varrho^4 (k^2 + m_\varrho^2)^{-2} (\mathrm{d}\sigma_\parallel/\mathrm{d}t - \mathrm{d}\sigma_\perp/\mathrm{d}t)_{\gamma N \to \pi \varDelta}. \tag{48}$$

Weaker crossing assumptions than in the case of photoproduction, namely for ratios of density matrix elements $\varrho^{00}/\varrho^{11}$ and $\varrho^{10}/\varrho^{11}$ only, are necessary also for the longitudinal parts of the electroproduction cross section if these are predicted [67] according to

$$\sigma_L(W^2, t, k^2)_{\pi^\pm} = \frac{k^2}{m_\varrho^2} \frac{m_\varrho^4}{(k^2 + m_\varrho^2)^2} (\varrho \varrho^{00}/\varrho \varrho_{11}) \tfrac{1}{2}(\mathrm{d}\sigma/\mathrm{d}t|_{\pi^+} + \mathrm{d}\sigma/\mathrm{d}t|_{\pi^-})$$
$$\pm (\varrho^0 \omega) \text{ interfer.} \tag{49}$$

For a discussion of the $\varrho^0 \omega$ interference term we refer to Ref. [67]. Its contribution to σ_L, and to σ_I, σ_T according to subsequent formulas has been neglected in the numerical analysis to be presented [67]. $\pi^\pm$ on the left hand side indicates whether π^+ or π^- is produced, i.e.

$$\gamma_{\text{virt}} \, p \to \pi^+ \varDelta^0 \qquad \pi^- p \to (\varrho^0, \omega) \, \varDelta^0$$
$$\gamma_{\text{virt}} \, n \to \pi^- \varDelta^+ \qquad \pi^+ n \to (\varrho^0, \omega) \, \varDelta^+ \tag{50}$$

or

$$\gamma_{\text{virt}} \, n \to \pi^+ \varDelta^- \qquad \pi^- p \to (\varrho^0, \omega) \, \varDelta^-$$
$$\gamma_{\text{virt}} \, p \to \pi^- \varDelta^{++} \qquad \pi^+ p \to (\varrho^0, \omega) \, \varDelta^{++} \tag{51}$$

and the ϱ^0 density matrix elements have to be taken from (50) or (51), corresponding to whether the left hand side in (49) refers to (50) or (51), $\mathrm{d}\sigma/\mathrm{d}t|_{\pi^\pm}$ denotes the corresponding real photoproduction cross section. Quite similar to (49) we have [67] for σ_I

$$\sigma_I(W^2, t, k^2) = -\sqrt{\frac{k^2}{m_\varrho^2}} \frac{m_\varrho^4}{(k^2 + m_\varrho^2)^2} [\varrho \varrho^{xz}/\varrho \varrho_{11}) \tfrac{1}{2}(\mathrm{d}\sigma/\mathrm{d}t|_{\pi^+} + \mathrm{d}\sigma/\mathrm{d}t|_{\pi^-})$$
$$\pm (\varrho^0 \omega) \text{ interfer.}]. \tag{52}$$

From (49) and the dominance of the production of longitudinally polarized ϱ^0 mesons for small $|t|$ in $\pi N \to \varrho^0 \varDelta$ [68, 69], we expect a dominating longitudinal electroproduction cross section for $k^2 > m_\varrho^2$ and for small $|t| \approx m_\pi^2$, quite similar to single pion electroproduction. If the energy W is not large enough such small values of $|t|$ will not be reached, ($|t_{\min}|$ is quite large, e.g. for $W^2 = 5.5 \text{ GeV}^2$, $k^2 = 0.5 \text{ GeV}^2/c^2$, $|t_{\min}| \approx 0.08 \text{ GeV}^2/c^2$), and because of the steep t dependence of $\pi N \to \varrho_{\text{long}} \varDelta(1236)$, the longitudinal cross section will correspondingly be smaller.

As no polarized photoproduction data were available, the subsequently discussed predictions for σ_T have actually not been based on (48) but rather on

$$\sigma_T(W^2, t, k^2)_{\pi^\pm} = \frac{m_\varrho^4}{(k^2 + m_\varrho^2)^2} [(-\varrho_{1\,-1}/\varrho_{11}) \tfrac{1}{2}(d\sigma/dt|_{\pi^+} + d\sigma/dt|_{\pi^-})$$

$$\pm (\varrho^0 \omega) \text{ interfer.}] \,. \tag{53}$$

First data on $\pi \varDelta$ electroproduction have very recently appeared [70, 71]. A comparison with predictions [67] for

$$\gamma_{\text{virt}} \, p \rightarrow \pi^+ \varDelta^0 (1236) \tag{54}$$

according to (47), (49), (52), (53) is shown on Fig. 12a, b. Photoproduction data [72] entering the predictions have been scaled from 16 GeV/c to the present energy ($W^2 = 5.5 \text{ GeV}^2$) assuming an $(s - M^2)^{-2}$ behaviour.

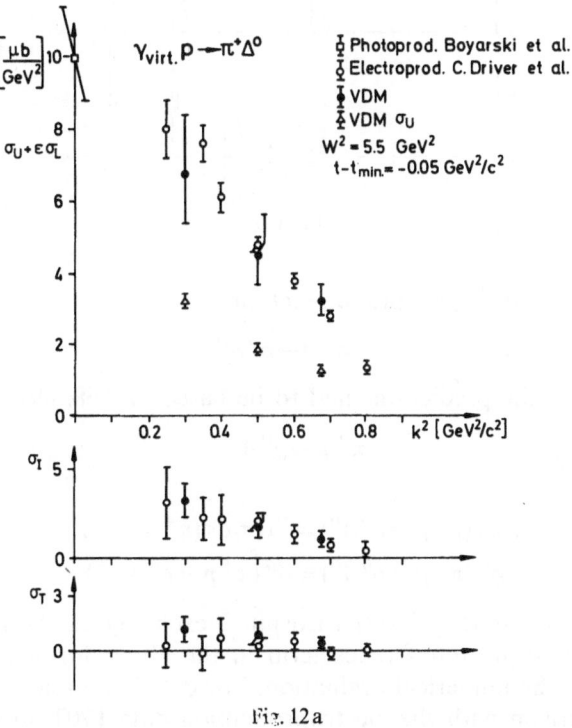

Fi?. 12a

Fig. 12. Comparison [67] of vector meson dominance predictions with electroproduction data [70] for $\gamma_{\text{virt}} p \rightarrow \pi^+ \varDelta^0$. a k^2 dependence at constant t, b t dependence

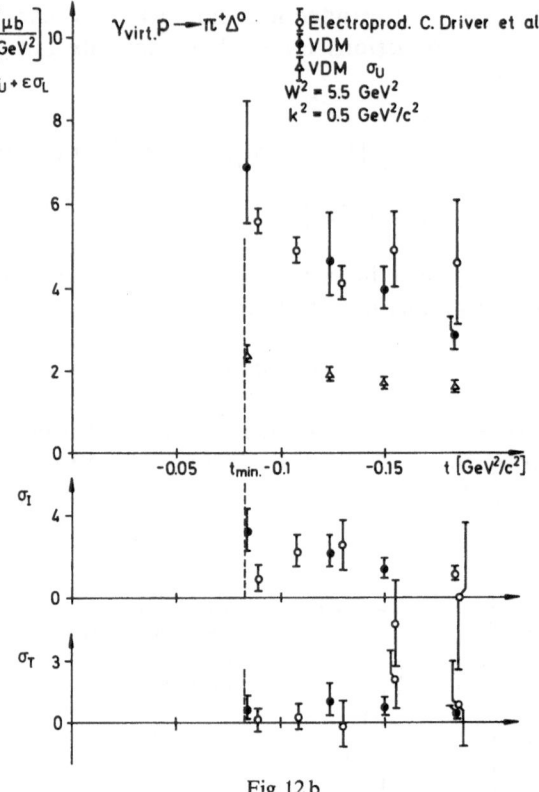

Fig. 12 b

As no data for the appropriate reaction

$$\pi^- p \to \varrho^0 \Delta^0 \tag{55}$$

are available, the predictions had to be based on [68, 69]

$$\pi^+ p \to \varrho^0 \Delta^{++} \tag{56}$$

using

$$d\sigma/dt(\pi^- p \to \varrho^0 \Delta^0) = \tfrac{1}{3} d\sigma/dt(\pi^+ p \to \varrho^0 \Delta^{++})$$
$$\varrho^{ik}(\pi^- p \to \varrho^0 \Delta^0) = \varrho^{ik}(\pi^+ p \to \varrho^0 \Delta^{++}) \tag{57}$$

which relations hold if (exotic) isospin 2 exchange in the t channel is negligible. The $\varrho\omega$ interference term in the above formulae has been neglected in the numerical evaluation. For details we refer to Ref. [67]. The comparison with the electroproduction data [70] on Figs. 12a, b shows good agreement [73]. For further discussion of $\pi\Delta$ electroproduction we also refer to Chapter 4.

3.4. A Brief Remark on the Total Longitudinal Electroproduction Cross Section

The electroproduction channels we have just discussed $\gamma_{\text{virt}} N \to \pi N$, $\pi \Delta (1236)$ are dominated by a large longitudinal contribution, which for single pion production is many times as large as the transverse cross section at $k^2 \cong m_\varrho^2$. This feature of dominating longitudinal contributions is in strong contrast to the behaviour of the total electroproduction cross section [74, 75], to which the longitudinal part only contributes roughly 20 %, as far as the transverse total cross section σ_T and the longitudinal one σ_L have been separated. In fact, all this relatively small longitudinal cross section [75]

$$\sigma_L \lesssim 12\,\mu\text{b} \quad \text{at} \quad W = 2.0\,\text{GeV}$$
$$k^2 = 0.5\,(\text{GeV}/c)^2$$

is almost saturated [76] by the few channels which have just been discussed:

$$\sigma_L(\gamma_{\text{virt}} p \to \pi^+ n, \pi^+ \Delta^0, \pi^- \Delta^{++}) \cong 7 \pm 2\,\mu\text{b} \quad [77].$$

Also $\gamma_{\text{virt}} p \to K^+ \Lambda, \Sigma$ shows [78] a k^2 behaviour typical for a large longitudinal contribution. It is, of course, an interesting question to answer experimentally, whether σ_L is given as a sum of these few channels also at higher energies, in which case σ_L is expected to go to zero for sufficiently high energy, or whether more general pion exchange dominated processes [79] will build up a σ_L constant in energy.

The fact that vector meson dominance is able to describe the different channels contributing to σ_L, as far as experimentally tested, fits in quite nicely with the recently proposed "two component picture" [80] (compare Chapter 5 for a brief discussion of it) for the total deep inelastic cross section. In this picture [80] the longitudinal part is assumed to be completely described by vector meson dominance, whereas the transverse part contains a vector meson dominance and a scale invariant contribution.

4. Vector Meson Dominance Predictions from Smoothness of Invariant Amplitudes

In Chapter 1.4 we have postulated the smoothness assumptions of vector meson dominance for s channel helicity amplitudes. In order to have a covariant starting point, and to completely separate dynamical and kinematical features, it seems advantageous to formulate [10, 15, 81] the smoothness assumption for the mass extrapolation in terms of invariant amplitudes, chosen to be free from kinematical singularities. This approach has been discussed most extensively for single pion production ($\gamma N \rightarrow \pi N$ and $\pi N \rightarrow \varrho^0 N$), and not only led to the same mass extrapolation procedure as it was formulated in 1.4 for the s channel helicity amplitudes at large energies, but moreover gave nontrivial additional relations [13, 82–85] between longitudinal and transverse amplitudes: The longitudinal amplitudes in fact turn out to be calculable from the transverse ones.

Let us only briefly show how the longitudinal transverse connection [13, 82] comes about for $VN \rightarrow \pi N$. One starts by decomposing the amplitude $T_\mu(VN \rightarrow \pi N)$ in terms of the well known 8 Ball [16] invariant amplitudes, which are known to be free from kinematic singularities

$$T_\mu(VN \rightarrow \pi N) = \sum_{i=1}^{8} M_\mu^{(i)} B_i (k^2 = -m_V^2, W^2, t). \tag{58}$$

Current conservation yields constraints, which for large energy $W^2 \equiv s$ have the form

$$\begin{aligned}
sB_2 - (t - m_\pi^2) B_3 &= -k^2(2B_4 + B_1) \\
-2B_5 + sB_6 - (t - m_\pi^2) B_8 &= -k^2(2B_7 + B_8).
\end{aligned} \tag{59}$$

On the vector meson mass shell $k^2 = -m_V^2$, these constraints are quite irrelevant, as they only define two additional amplitudes which do not enter the 6 helicity amplitudes calculated from (58). (59) is important, however, for the extrapolation in k^2 and for $k^2 = 0$ correctly reduces the number of invariant amplitudes to 4. If independence of k^2 is required for all 8 invariant amplitudes for large W^2, the left hand side in (59) has to vanish for arbitrary k^2, as it vanishes for photons, and moreover the combination of invariant amplitudes on the right hand side of (59) has to vanish, thus yielding 4 conditions between 8 amplitudes. Only 4 independent amplitudes remain thus allowing to express longitudinal amplitudes in terms of transverse ones [13, 82, 83], the explicit relation being [13]:

$$\begin{aligned}
f_{1,01}^{(VN \rightarrow \pi N)} &= \tfrac{1}{2} m_V (-2t)^{-1/2} (f_{1,11}^{VN \rightarrow \pi N} - f_{1,-11}^{VN \rightarrow \pi N}) \\
f_{-1,01}^{(VN \rightarrow \pi N)} &= -m_V (-2t)^{-1/2} f_{-1,11}^{VN \rightarrow \pi N}.
\end{aligned} \tag{60}$$

The smoothness assumption for the amplitudes B_i may be verified to hold [13] in the high energy limit for simple Born term models.

The consequences from (60) are twofold: On the one hand relations between different density matrix elements for $\pi N \rightarrow \varrho^0 N$ may be obtained and compared [87] with experiment, on the other hand the transverse amplitudes in (60) may be related to photoproduction, thus predicting transverse and longitudinal ϱ^0 production from photoproduction. Moreover in a straightforward manner, the procedure may be generalized [13, 59] to electroproduction, thus predicting transverse and longitudinal electroproduction via vector meson dominance, current conservation and smooth invariant amplitudes. As no complete set of photoproduction experiments determining the invariant amplitudes uniquely is available, the procedure is to make a model parameterization for photoproduction and then predict the vector meson induced reaction and electroproduction.

Let us stress at this point again that the formulation of the mass extrapolation of vector meson dominance in terms of smooth invariant amplitudes is completely consistent for single pion production with the formulation in terms of s channel helicity amplitudes of Chapter 1.4: Starting from smoothness of the Ball invariant amplitudes, which is fulfilled in the electric Born term model, in the high energy limit the transverse s channel helicity amplitudes are mass independent, whereas the longitudinal ones have a factor $\sqrt{-k^2}$ as in (27). An analogous result has been obtained for $\pi\Delta(1236)$ production, which will be briefly discussed below.

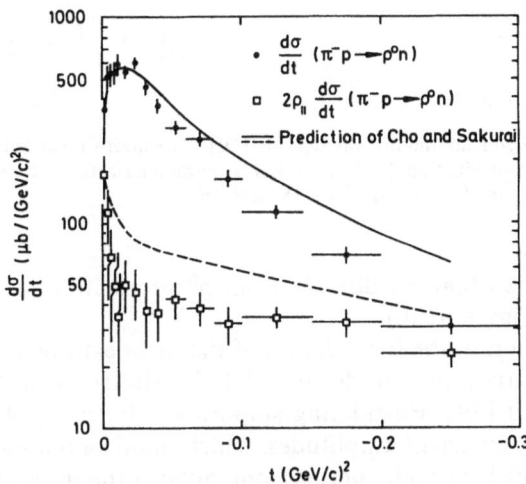

Fig. 13. Vector meson dominance predictions [13] for $\pi^- p \to \varrho^0 n$ from photoproduction via smooth invariant amplitudes. The data are from Ref. [41]

Fig. 13 shows the predictions [13] thus obtained for ϱ^0 production from a finite energy sum rule parameterization [88] of photoproduction. The prediction for the transverse ϱ^0 production lies somewhat high in comparison with the ϱ^0 data, a feature which, of course, has also been observed in the direct comparison as shown on Fig. 6 and discussed in Chapter 3.1. For similar predictions from a Regge pole parameterization of photoproduction, we refer to Ref. 83.

A comparison with experiment [49] of predictions obtained [59] with this procedure for single pion electroproduction from a Regge parameterization of photoproduction is shown on Figs. 14a, b and shows quite good agreement also for σ_I. The results should be compared

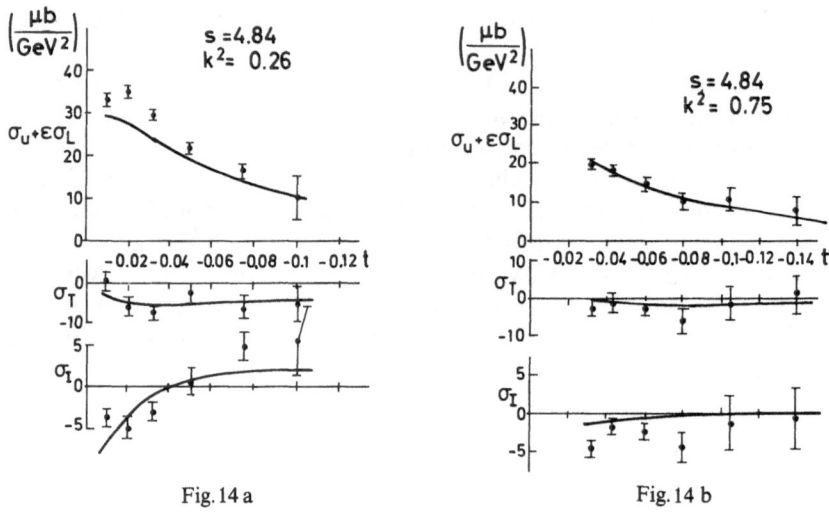

Fig. 14 a   Fig. 14 b

Fig. 14. Vector meson dominance predictions [59] via smooth invariant amplitudes for single pion electroproduction [49] from Regge parameterization of photoproduction. a for $k^2 = +0.026 \text{ GeV}^2/c^2$, b for $k^2 = 0.75 \text{ GeV}^2/c^2$

with the results obtained directly from ϱ^0 production as discussed in Chapter 4.2 (Figs. 8 to 10).

The extension of the formulation of the mass extrapolation in terms of smooth invariant amplitudes to $\pi\varDelta(1236)$ electroproduction has just been completed [89]. Postulating smoothness for a set of kinematical singularity free invariant amplitudes, which fulfill smoothness in simple Born term models at high energies, one obtains the mass extrapolation behaviour (27) for the s channel helicity amplitudes at high energies. In order to make predictions for electroproduction from photoproduction, photoproduction has been described [90] by the electric Born term model [66] with absorption corrections [91]. The result obtained is shown on Fig. 15, and the fit to photoproduction, which has been used as input, on Fig. 16. There are discrepancies for electroproduction especially, apparently for the longitudinal part, which seems to be too small. In view of the rather encouraging better agreement obtained in

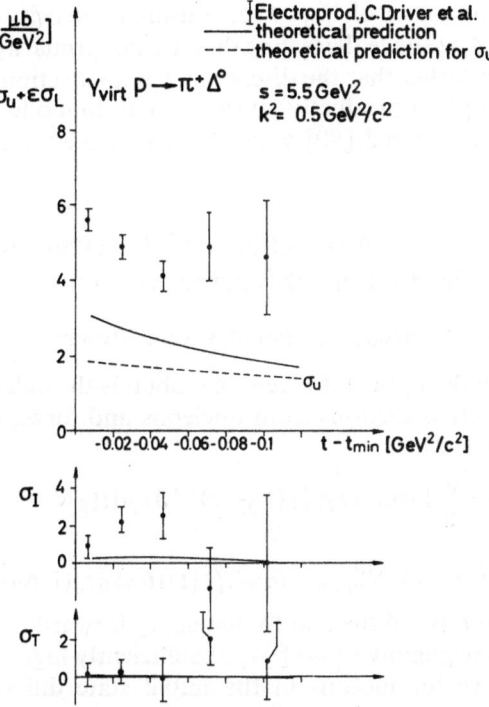

Fig. 15. Vector meson dominance predictions [89] for $\pi\Delta(1236)$ electroproduction [70] using smoothness of invariant amplitudes and a parametrization [90] of photoproduction in terms of a Born term model with absorption corrections

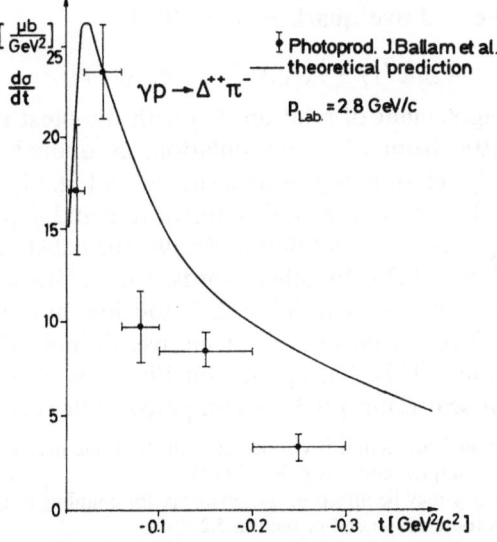

Fig. 16. Theoretical description [90] of $\pi\Delta(1236)$ photoproduction, Born term model with absorption corrections, used as input for the electroproduction predictions of Fig. 15

the direct comparison of Figs. 12a, b using ϱ^0 density matrix elements from $\pi N \to \varrho^0 \Delta$, we have the feeling that it is not primarily the smoothness that failed, but rather that the discrepancy is a reflection of discrepancies also present in photoproduction if the Born term model with absorption corrections is compared [90] with the density matrix elements.

5. The Total Photoproduction and Electroproduction Cross Section and Production of Vector Meson

5.1. Photo- and Electroproduction of Vector Mesons

From (31) with the optical theorem one obtains the well known relations between total cross sections from nucleons and forward production of vector mesons [92]

$$\sigma_{\text{tot}}(\gamma N) = \sum_V \left[16 \pi (\alpha \pi / \gamma_V^2) (1 + \alpha_V^2)^{-1} d\sigma/dt(\gamma N \to V N)|_{t=0} \right]^{1/2} \quad (61)$$

and [93]

$$d\sigma/dt(\gamma N \to V N)|_{t=0} = (\alpha \pi / \gamma_V^2) (1/16\pi) \sigma_{\text{tot}}^2(V N) (1 + \alpha_V^2), \quad (62)$$

where α_V is the ratio of the real to imaginary forward production amplitude and α_V^2 is negligibly small [94] at sufficiently high energies. Contributions from vector mesons in the initial state differing from V are assumed to be small in (62). A third relation, between $\sigma_{\text{tot}}(\gamma N)$ and $\sigma_{\text{tot}}(V N)$ follows immediately by substituting (62) into (61). A very recent comparison [95] of (62) for forward ϱ^0 production [96–101] from protons extrapolated to $t = 0$ is shown on Fig. 17, taking the total $\varrho^0 p$ cross section from the additive quark model [102][3]

$$\sigma_{\text{tot}}(\varrho^0 p) = \tfrac{1}{2}(\sigma_{\text{tot}}(\pi^+ p) + \sigma_{\text{tot}}(\pi^- p)). \quad (63)$$

There is good agreement of relation (62) with the most recent value of $\gamma_\varrho^2/4\pi = 0.64 \pm 0.06$ from $e^+ e^-$ annihilation, as quoted in Chapter 2. In order to satisfy relation (61), as also shown on Fig. 17, when inserting the measured values of $\sigma_{\text{tot}}(\gamma p)$ and of forward ω and ϕ production and the measured γ_ϕ, $\gamma_\omega e^+ e^-$ annihilation results, the constant $\gamma_\varrho^2/4\pi$ comes out too low $\gamma_\varrho^2/4\pi = 0.35$. In other words, with $\gamma_\varrho^2/4\pi = 0.64$, the prediction for $\sigma_{\text{tot}}(\gamma p)$ becomes roughly 25% too low in comparison with experiment: the 3 vector mesons by themselves do not fully saturate the total cross section[4]. This discrepancy on Fig. 18 is also seen for $t \neq 0$, where Compton scattering [103] is compared with vector meson pro-

---

[3] The quark model prediction is consistent with the value measured for $\sigma_{\text{tot}}(\varrho^0 p)$ in ϱ^0 production from complex nuclei (e.g. Ref. [119]).

[4] This discrepancy may be interpreted as evidence for coupling of higher mass vector states. Compare footnote at the end of section 5.2.

duction, testing essentially the square of (61), but also for $t \neq 0$ (using $\gamma_\varrho^2/4\pi = 0.5$):

$$d\sigma/dt(\gamma p \rightarrow \gamma p) = \left(\sum_{V = \varrho^0, \omega, \phi} (\alpha\pi\gamma_V^{-2} d\sigma^{tr}/dt(\gamma p \rightarrow Vp)^{1/2}\right)^2. \quad (64)$$

The data [103] are consistent with no change in slope, when extrapolating from the outgoing vector meson mass to the outgoing $k^2 = 0$ photon.

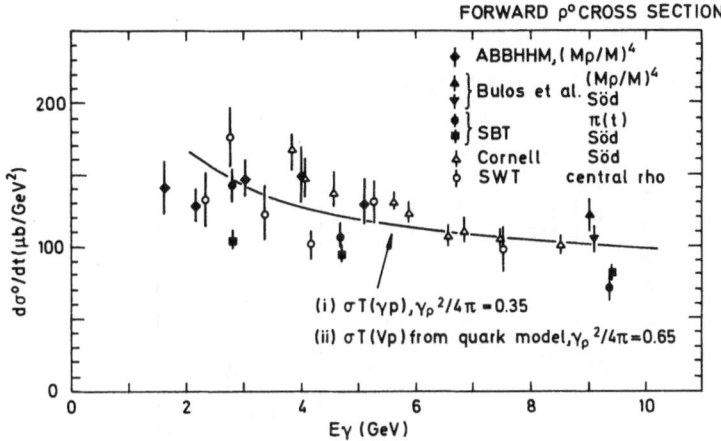

Fig. 17. Comparison of forward ϱ^0 production [96 to 101] (extrapolated to $t = 0$) with a fit to the measured $\sigma_{tot}(\gamma p)$ according to (61) and with the vector meson dominance quark model relation (62), (63) (from Ref. [95]

From the good agreement of (62) for forward ϱ^0 production, one would expect smoothness to be well fulfilled for $\varrho^0 p \rightarrow \varrho^0 p$, when extrapolating further also into the spacelike region as investigated in electroproduction $\gamma_{virt} p \rightarrow \varrho^0 p$ [104]. First measurements have recently been completed at DESY [105], Cornell [106] and SLAC [107]. The DESY apparatus measures directly the decay pions, with a limited acceptance, however, such that transversely polarized ϱ^0 mesons are detected only. Assuming helicity-conservation, which is well fulfilled for photoproduction (see e.g. the review Ref. [95]), what is thus measured is the term $\sigma_u \equiv d\sigma_u/dt$ in the general electroproduction formula (39). From vector meson dominance we expect

$$\frac{(k^2 + m_\varrho^2)^2}{m_\varrho^4} d\sigma_u/dt(\gamma_{virt} p \rightarrow \varrho^0 p) = d\sigma/dt(\gamma p \rightarrow \varrho^0 p)$$

$$= (\alpha\pi/\gamma_\varrho^2) d\sigma/dt(\varrho^0 p \rightarrow \varrho^0 p). \quad (65)$$

i.e. constancy of the transverse ϱ^0 electroproduction cross section after multiplication with the ϱ^0 propagator factor. From Fig. 19 we see that

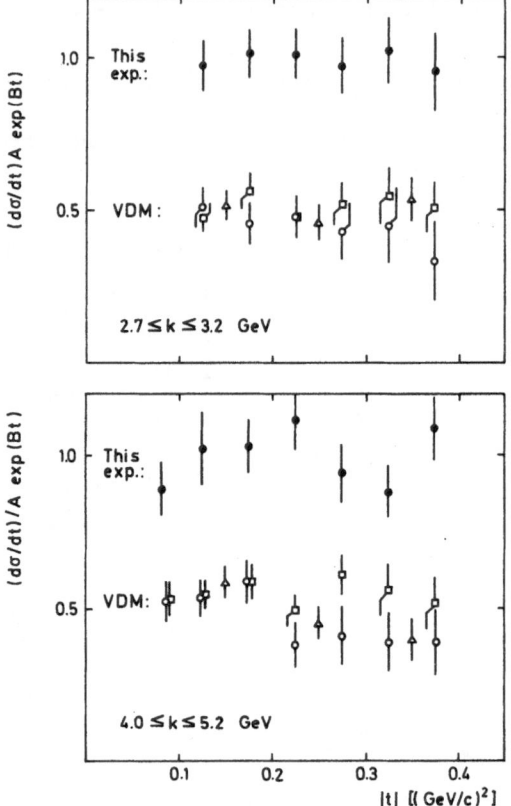

Fig. 18. Comparison of vector meson production with Compton scattering according to (64) ($\gamma_\varrho^2/4\pi = 0.5$) (From Ref. [103])

the ϱ^0 electroproduction data [105] lie significantly higher than expected, showing a k^2 behaviour more proportional to the total transverse electroproduction [74, 75] cross section as indicated. As we are comparing $\pi p \to \pi p$ data [108] with photo- [96–100] and electroproduction [105] outside the forward direction $|t| = 0.08$ GeV$^2/c^2$ one possible explanation of the behaviour shown in Fig. 19 would be a flatter t behaviour [109–111] for electroproduction, but data [105] seem to be consistent with no change of slope. Indications of a change in slope have been observed in the SLAC [107] and Cornell [106] experiments, but as the recoil proton and not the $\pi^+\pi^-$ system is measured, because of background problems, no unambiguous conclusion can be drawn on the ϱ^0 slope so far. We can look forward with great interest to further results on ϱ^0 electroproduction, which are expected to come out in the near future at the different electron laboratories.

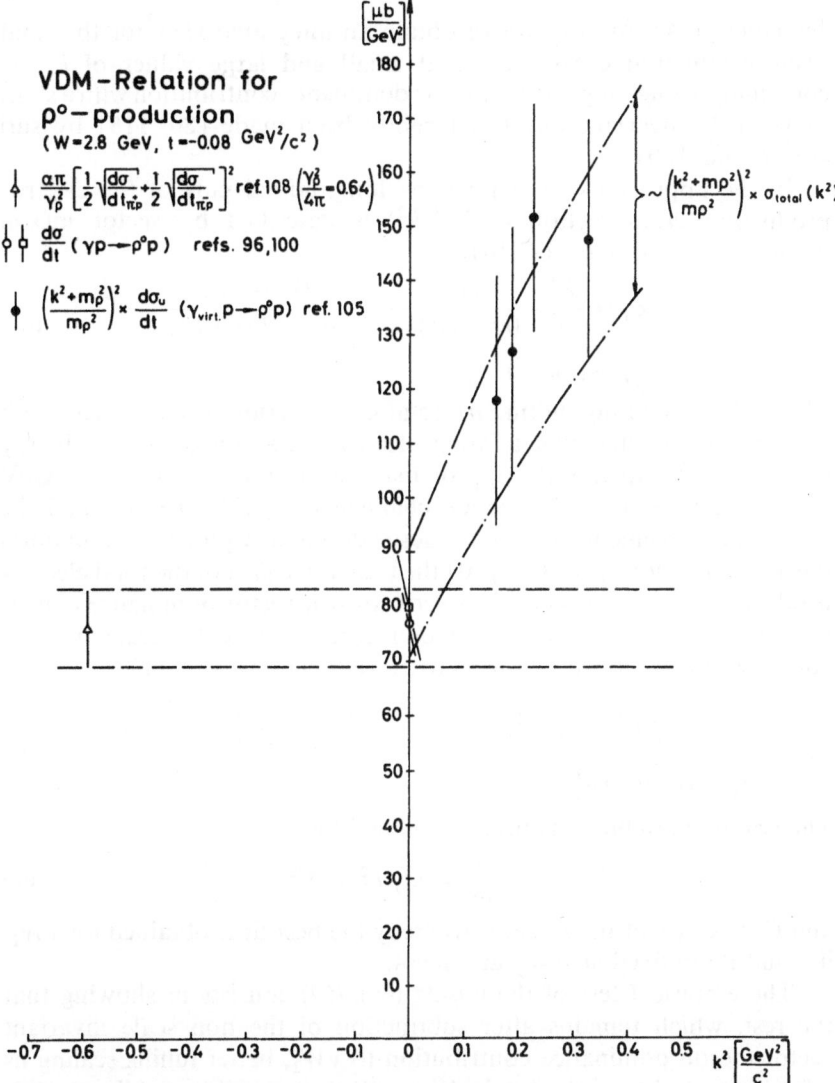

Fig. 19. Comparison of ϱ^0 scattering at $t = -0.08$ GeV$^2/c^2$ according to the quark model relation with ϱ^0 production by real and virtual photons

5.2. The Total Electroproduction Cross Section

As is well known by now, vector meson dominance fails [112] for the total electroproduction cross section for large spacelike k^2, and the large transverse cross section observed led to the parton [113] picture and to attempts to understand the observed scaling behaviour [112] in terms of electromagnetic current commutator expansions near the lightcone [114]. An attempt to obtain an integrated view for the total electroproduction cross section at small and large values of k^2 by combining a long range vector meson dominance contribution with a short range scale invariant one has recently been made [80, 115] by suri and Yennie [80].

In the approach of Ref. [80] the longitudinal part of the electroproduction cross section $\sigma_L(W^2, k^2)$ is described by vector meson dominance according to [116]

$$\sigma_L = \frac{k^2}{m_\varrho^2} \frac{m_\varrho^4}{(k^2 + m_\varrho^2)^2} \sigma_{\gamma L} \frac{W^2 - M^2}{W^2 - M^2 + k^2} \qquad (66)$$

$$\sigma_{\gamma L} \equiv (\alpha \pi / \gamma_\varrho^2) \sigma_{\varrho p}^L$$

where $\sigma_{\varrho p}^L$ is the longitudinal ϱp total cross section. $\sigma_{\gamma L}$ is actually left as a free parameter, when fitting electroproduction data. As already remarked in Chapter 3, the approximate saturation of σ_L at $W \cong 2\,\mathrm{GeV}$ by a few two particle production channels as $\gamma_{\mathrm{virt}} N \to \pi N, \pi \varDelta$, and the reasonable agreement with vector meson dominance predictions supports the ansatz (66) for σ_L. In Ref. [80] the transverse part of the total electroproduction cross section σ_T is described by a vector dominance contribution plus a rather ad hoc ansatz for a term which yields a scaleinvariant contribution to the well known structure function νW_2 [117].

$$\sigma_T = m_\varrho^4 (k^2 + m_\varrho^2)^{-2} \sigma_{\gamma T} + \frac{2MG x'^{a-2}(1 - x')^\beta}{(1 - 2Bx' + Cx'^2)(W^2 - M^2)}, \qquad (67)$$

$$\sigma_{\gamma T} \equiv (\alpha \pi / \gamma_\varrho^2) \sigma_{\varrho p}^T.$$

The scaling variable x' is used as defined by

$$x' = \frac{1}{\omega'} = k^2/(k^2 + W^2) \qquad (68)$$

and G, B, C are fit parameters, whereby the best fit is obtained for νW_2, if a and β are fixed as $a = 2$ and $\beta = 4$.

The empirical test of the ansatz (66), (67) consists in showing that the rest, which remains after subtraction of the non scale invariant vector meson dominance contribution to νW_2, better fulfills scaling as a function of x' than the total $\nu W_2 \equiv \nu W_2(\mathrm{tot})$, which according to (66) and (67) consists of a sum of terms:

$$\nu W_2(\mathrm{tot}) = \nu W_2(VT) + \nu W_2(VL) + \nu W_2(SC) \qquad (69)$$

In this equation, $\nu W_2(VT)$ and $\nu W_2(VL)$ denote the contributions to νW_2 stemming from vector meson dominance, whereas $\nu W_2(SC)$ denotes the scale invariant rest. Fig. 20, in which all available data points [112] for νW_2 have been plotted, indeed shows that $\nu W_2(SC)$ i.e. the rest after subtracting vector meson dominance contributions, scales better than $\nu W_2(\mathrm{tot})$: In the region of small x' (large ω') where vector meson contributions are expected to be important, the data for $\nu W_2(SC)$ lie much

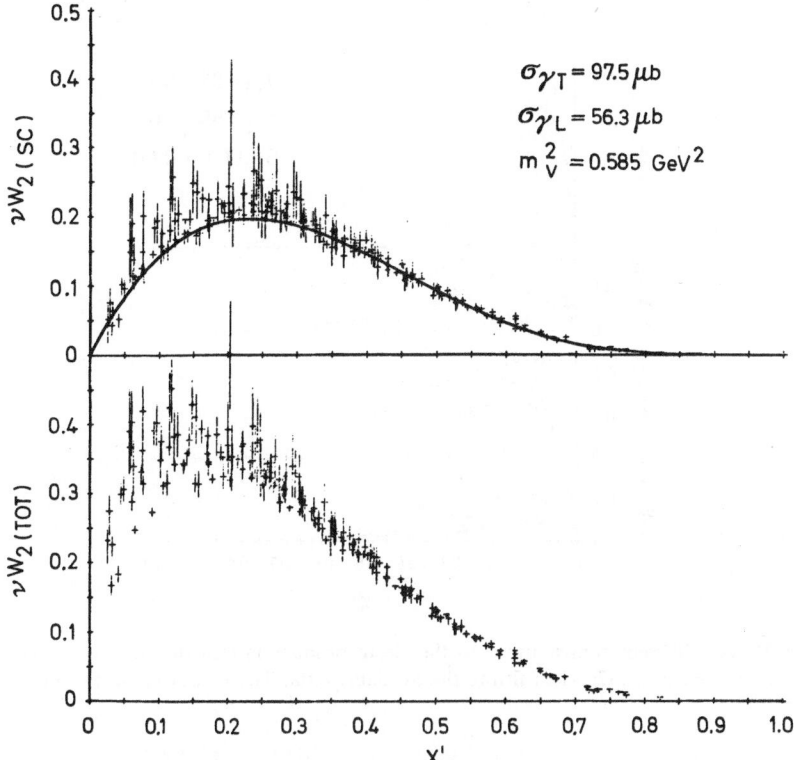

Fig. 20. The electroproduction structure function $\nu W_2 \equiv \nu W_2$ (tot) as a function of x', and νW_2(SC), which has been obtained from νW_2 after subtracting a vector meson dominance contribution according to (69), (67), (66) (From Ref. [80])

closer together on a universal curve than the data for νW_2(tot). The value obtained from the fit for $\sigma_{\gamma T} = 98 \pm 6\,\mu$b lies reasonably close to the asymptotic value for the total real photoproduction cross section, which from fits to the data has been obtained to be [80] $\sigma_{\gamma T} = 108 \pm 2\,\mu$b. The ratio of longitudinal to transverse total ϱ^0 cross sections is obtained unequal 1, whereas perhaps one would have preferred a value 1 consistent with spin independence at high energies. Finally on Fig. 21 one sees the curves as they are obtained [80] from the fit for the different contributions to νW_2(tot). Measurements at higher e.g. NAL energies to measure the behaviour of νW_2 for small x' will of course be most important to more clearly test such a "two component" picture, which in our opinion constitutes a first step for obtaining a more unified view of the electromagnetic interactions of the hadrons, as they manifest themselves in the behaviour of the total electroproduction cross section[5].

---

[5] *Note added in proof:* More recently it has been shown [120] that by taking into account the coupling of the photon to higher mass vector states σ_T may be successfully predicted for large ω' with essentially no adjustable parameter. Compare Fig. 22. The model also successfully predicts the measured [112] n/p ratio [120].

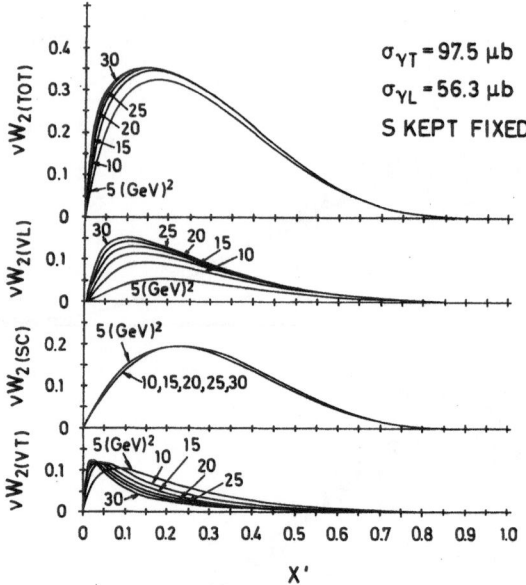

Fig. 21. The different contributions to the electroproduction structure function νW_2, as obtained from (66), (67) when fitting the available data. The parameter in the Fig. is s $(s \equiv W^2)$

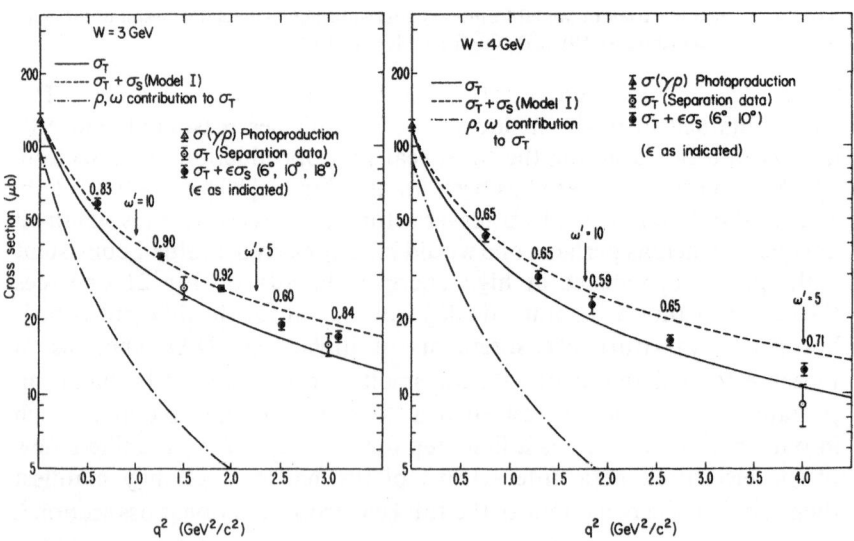

Fig. 22. Total virtual photon-proton cross sections σ and $\sigma_T + \sigma_S$ as functions of the four momentum squared q^2 of the virtual photon for (a) $W = 3$ GeV and (b) $W = 4$ GeV (from Ref. [120])

6. A Few Summarizing and Concluding Remarks

1. The main progress during the past year related to vector meson dominance, in our opinion, has been made through the advent of data on quasi two-body electroproduction beyond the resonance region. These data showed that the simple mass extrapolation assumptions for the longitudinal amplitude, mainly based on the $\sqrt{\pm k^2}$ factor originating from current conservation, are in fact able to reproduce the data quite well for the few channels $\gamma_{\mathrm{virt}} N \to \pi N,, \pi \Delta, K \Lambda$, which are dominated by a large longitudinal contribution and at $W \cong 2\,\mathrm{GeV}$ roughly build up the total longitudinal electroproduction cross section. These data thus support the ansatz of suri and Yennie for the total electroproduction cross section, in which the longitudinal part consists of a vector meson dominance contribution only, whereas the transverse part also contains a short range, scale invariant part. Further measurements of the above mentioned pseudoscalar meson production channels at larger energies and photon four momentum squared will, of course, be most important for further refined tests of the model. The same holds for the total electroproduction cross section, where further exploration of the large ω' region with σ_T, σ_L separation will be extremely important.

2. Whereas ϱ^0 photoproduction is at present in good agreement with predictions from ϱ^0 scattering (e.g. via πN scattering and quark model), first measuments on ϱ_0 electroproduction indicate deviations (Fig. 19). Further experimental work, also to clarify questions like k^2 dependence of the ϱ^0 slope, will of course be most important and is under preparation in most electron laboratories.

3. Formulating vector meson dominance in terms of smooth invariant amplitudes chosen to be free from kinematic singularities and to fulfil smoothness in simple dynamical models at high energies, allows a rather complete separation of kinematical and dynamical features. Maximal similarity between real and virtual photons and massive vector mesons is thus obtained: The longitudinal helicity amplitudes are related to the transverse ones and the same number of independent helicity amplitudes governs the behaviour of massive and massless vector particles. Agreement with experiment in the overall features is quite good for single pion photo- and electroproduction and ϱ^0 production by pions. Further experimental and theoretical work is especially needed for $\pi \Delta$ production. From the theoretical side, questions as e.g. the uniqueness or non uniqueness of the choice of the invariant amplitudes and their behaviour in dynamical models will be of interest.

4. Future results on $e^+ e^-$ annihilation in the years to come will undoubtedly settle questions like the existence of higher vector mesons, and to what extent the production mechanism at high energies is dominated by production via the tails of the known vector mesons.

Acknowledgement. Discussions on the DESY electroproduction data with Dr. K. Heinloth are gratefully acknowledged. Thanks are also due to my theoretical collegues at DESY for useful discussions.

Footnotes and References

1. Donnachie, A.: Lecture given at the Hamburg Summerschool 1971 (this volume, p. 121) and Proc. of the 1971 Internat. Symp. on Electron and Photon Interactions at High Energies Cornell University, Ithaca, N.Y. (1971).

2. Erwin, A., *et al.:* Phys. Rev. Letters **6**, 628 (1961). – Stonehill, D., *et al.*: Phys. Rev. Letters **6**, 624 (1961). – Bertansa, L., *et al.*: Phys. Rev. Letters **9**, 180 (1962). – Maglic, B.C., *et al.*: Phys. Rev. Letters **7**, 178 (1961); Phys. Rev. **125**, 687 (1962). – Xuong *et al.*: Phys. Rev. Letters **7**, 327 (1961).

3. e.g. Particle data group. Rev. Mod. Phys. **43**, 2 (1971).

4. e.g. Review talks by Perez-y-Jorba, J., Sidorov, V.A.: Proc. of the 4th International Symposium on Electron and Photon Interactions at High Energies, Liverpool (1969), and the reviews Lefrancois, J., Sidorov, V.A. in Proc. of the 1971 Internat. Symposium on Electron and Photon Interactions at High Energies, Cornell University, Ithaca, N.Y. (1971).

5. For earlier reviews on vector meson dominance we refer to e.g. Sakurai, J.J.: Proc. of the 4th Int. Symposium on Electron and Photon Interactions at High Energies, Liverpool 1969. – Joos, H.: Acta Physica Austriaca Suppl. IV 1967. – Schildknecht, D.: Z. Phys. **229**, 278 (1969) and DESY 69/41 publ. in Proc. of the Int. Seminar on Vector Mesons and Elrctromagnetic Interactions, Dubna 2-4816 (1969) and DESY 70/31, Proc. of the 5. Rencontre de Moriond (1970).

6. Cho, C.F., Sakurai, J.J.: Phys. Letters **31**B, 22 (1970). – Nieh, T.H.: Phys. Rev. D**1**, 3161 (1970).

7. Sakurai, J.J.: Phys. Rev. Letters **22**, 981 (1969).

8. Fraas, H., Schildknecht, D.: Nucl. Phys. B**14**, 543 (1969).

9. Iso, C., Yoshii, H.: Ann. Physik **51**, 490 (1969). – Iso, C., Schildknecht, D.: Nucl. Phys. B**21**, 242 (1970).

10. Fraas, H., Schildknecht, D.: Nucl. Phys. B**6**, 395 (1968).

11. Diederich, S.: Private communication reported in Schildknecht, D., review given at Dubna, see Ref. [5].

12. Cho, C.F., Sakurai, J.J.: Phys. Letters **30**B, 119 (1969).

13. — — Phys. Rev. D**2**, 517 (1970). – Cho, C.F.: Phys. Rev. D**4**, 194 (1971).

14. Kellett, B.H.: Nucl. Phys B **35**, 517 (1971).

15. Le Bellac, M., Plaut, G.: Nuovo Cimento **64**A, 95 (1969).

16. Ball, J.S.: Phys. Rev. **124**, 2014 (1961).

17. Nambu, Y.: Phys. Rev. **106**, 1366 (1957).

18. Frazer, W.R., Fulco, J.R.: Phys. Rev. Letters **2**, 365 (1959) and Phys. Rev. **117**, 1603 (1960).

19. Sakurai, J.J.: Ann. Physik **11**, 1 (1960).

20. Joos, H.: Ref. [5].

21. Kroll, N.M., Lee, T.D., Zumino, B.: Phys. Rev. **157**, 1376 (1967).

22. Gell-Mann, M., Zachariasen, F.: Phys. Rev. **124**, 953 (1961). — Gell-Mann, M.: Phys. Rev. **125**, 1067 (1962).

23. See the review by Lefrancois, J., Ref. [4].

24. See e.g. the review by Renard, F.M., Lectures at the International Summer Institute of Theoretical Physics, Hamburg (1971). This volume, p. 98

25. Gournaris, G., Sakurai, J.J.: Phys. Rev. Letters **21**, 244 (1968).

26. See Perez-y-Jorba, J.: Ref. [4].
27. Bernardini, C.: in Proc. of the 1971 Internat. Symposium on Electron and Photon Interactions at High Energies. Cornell University, Ithaca, N.Y. (1971).
28. Kramer, G., Uretsky, J. L., Walsh, T. F.: Phys. Rev. D3, 719 (1971) and private communication.
29. Layssac, J., Renard, F. M.: Il Nuovo Cimento 6A, 134 (1971) and Letters Nuovo Cimento 1, 197 (1971).
30. See also Bramon, A., Greco, M.: Int. Rep. 71/8 Frascati (1971). – Vaughn, M. T., Polito, P. J.: Lett. Nuovo Cimento 1, 74 (1971). – Altukhov, A. M., Khriplovich, I. B.: Preprint Novosibirsk 1971.
31. Beder, D. S.: Phys. Rev. 149, 1203 (1966).
32. Krammer, M., Schildknecht, D.: Nucl. Phys. B7, 583 (1968).
33. Iso, C., Yoshii, H.: Ann. Physik 47, 424 (1968).
34. Besides Refs. [31 to 33 and 35] see Krammer, M.: Phys. Letters 26B, 633 (1968) Erratum Phys. Letters 27B, 260 (1968). – Dar, A., Weisskopf, V. F., Levinson, C. A., Lipkin, H. J.: Phys. Rev. Letters 20, 1261 (1968). – Diebold, R., Poirier, J. A.: Phys. Rev. Letters 20, 1532 (1968). – Derado, I., Guiragossian, Z. G. T.: Phys. Rev. Letters 21, 1556 (1968). – Diebold, R., Poirier, J. A.: 22, 255 (1969); 22, 906 (1969).
35. Geweniger, C., et al.: Phys. Letters 28 B, 155 (1968). – Burfeindt, H., et al.: Phys. Letters 33 B, 509 (1970) and abstract 86, Liverpool Conf. (1969). – Geweniger, C., et al.: Phys. Letters 29 B, 41 (1969).
36. Bialas, A., Zalewski, K.: Phys. Letters 28B, 439 (1969).
37. Donohue, J. T., Högaasen, H.: Phys. Letters 25B, 554 (1967).
38. Besides Refs. [10, 11, 12, 15] we refer to Potter, W. T., Sullivan, J. D.: Nuovo Cimento 68A, 623 (1970). – Schmidt, W.: Phys. Rev. 188, 2458 (1969). – Brown, S. G.: Phys. Rev. D1, 207 (1970). – Dar, A.: Nucl. Phys. B19, 259 (1970). – Donohue, J. T.: Phys. Rev. D1, 1972 (1970). – Eilam, G., Berlad, G., Dar, A.: Nucl. Phys. B27, 415 (1971). – Berlad, G., Eilam, G.: Nucl. Phys. B25, 321 (1971).
39. Schildknecht, D.: DESY 69/41, also quoted under Ref. [5].
40. Dar, A.: Ann. Phys. 65, 324 (1971).
41. Bulos, F., et al.: Phys. Rev. Letters 26, 1457 (1971).
42. — Phys. Rev. Letters 26, 1453 (1971).
43. Boyarski, A. M., et al.: Phys. Rev. Letters 20, 300 (1968). – Heide, P., et al.: Phys. Rev. Letters 21, 248 (1968).
44. e.g. Richter, B., Harari, H.: Proc. of the International Symposium on Electron and Photon Interactions at High Energies, Stanford 1967.
45. Boyarski, A. M., et al.: Ref. [43] and Phys. Rev. Letters 21, 1767 (1968).
46. Schmidt, W.: Ref. [38]. – Manweiler, R., Schmidt, W.: Phys. Letters 33B, 366 (1970) and Phys. Rev. D 3, 2752 (1971).
47. Koch, W.: Private communication.
48. Brown, C. N., et al.: Phys. Rev. Letters 26, 987 and 991 (1971).
49. Driver, C., et al.: Phys. Letters 35B, 77, 81 (1971).
50. Kummer, P. S., et al.: DNPL/p67 (1971).
51. e.g. Berkelmann, K.: Phys. Rev. Letters 14, 1036 (1965).
52. Sullivan, J. D.: Phys. Letters 33B, 179 (1970).
53. Cho, C. F.: Phys. Rev. D4, 194 (1971).
54. Fraas, H., Schildknecht, D.: Phys. Letters 37 B, 389 (1971).
55. — — Letters 35B, 72 (1971).
56. Berends, F. A., Gastmans, R.: Phys. Rev. Letters 27, 124 (1971).
57. Miller, D. H., et al.: Phys. Rev. 153, 1923 (1967). – Johnson, P. B., et al.: Phys. Rev. 176, 1651 (1968). – Poirier, J. A., et al.: Phys. Rev. 163, 1462 (1967). – Hyams, B. D., et al.: Nucl. Phys. B7, 1 (1968). – Bulos, F., et al.: Phys. Rev. Letters 26, 1453 (1971).

58. Burfeindt, H., *et al.*: Ref. [35]. – Heide, P., *et al.*: Ref. [43].
59. Kellett, B. H.: Nucl. Phys. B **38**, 573 (1972).
60. Cho, C. F.: Phys. Rev. **5**, 260 (1972).
61. Schmidt, W.: DESY 71/22.
62. Berends, F. A.: Phys. Rev. D **1**, 2590 (1970). – Berends, F. A., Gastmans, R.: Phys. Rev. **5**, 204 (1972).
63. Devenish, R. C. E., Lyth, D. H.: Phys. Rev. **5**, 47 (1972).
64. Iso, C., Yoshii, H.: Ann. Phys. **48**, 237 (1968). – Dar, A.: Nucl. Phys. B **11**, 634 (1969).
65. Boyarski, A. M., *et al.*: Phys. Rev. Letters **25**, 695 (1970).
66. Stichel, P., Scholz, M.: Nuovo Cimento **34**, 1381 (1964).
67. Bartl, A., Schildknecht, D.: Nucl. Phys. B **36**, 28 (1972).
68. Brown, D. G.: University of California. Thesis, UCRL – 18254.
69. Aderholz, M., *et al.*: Nucl. Phys. B **22**, 1 (1970) and Nucl. Phys. B **24**, 509 (1970).
70. Driver, C., *et al.*: Nucl. Phys. B **32**, 45 (1971).
71. Brown, C. N., *et al.*: Harvard preprint (1971).
72. Boyarski, A. M., *et al.*: Phys. Rev. Letters **25**, 695 (1970).
73. A phase space factor has been neglected in Ref. [67], which when included will lower the predictions for $\sigma_u + \varepsilon\sigma_L$ by roughly 20%, which is inside the experimental errors.
74. Bloom, E. D., *et al.*: SLAC-PUB 796, SLAC-PUB-805.
75. Brasse, F. W., *et al.*: DESY 71/19.
76. Compare also the discussion given by Harari, H., Proc. of the 1971 International Symposium on Electron and Photon Interactions at High Energies, Cornell University, Ithaca, N. Y. (1971).
77. Heinloth, K.: Private communication. The value has been obtained from data of Refs. [49, 70, 71] by extrapolating the t behaviour observed at small t to t_{max}.
78. Brown, C. N., *et al.*: Phys. Rev. Letters **28**, 1086 (1972).
79. Drell, S. D.: Phys. Rev. Letters **5**, 278 (1961).
80. suri, A., Yennie, D. R.: SLAC-PUB-954, Ann. of Physics (to be published).
81. See also the literature given in 38.
82. Achasov, N. N., Shestakov, G. N.: J. Nucl. Phys. **11**, 1090 (1970).
83. Kellet, B. H.: Ref. [14] and Nucl. Phys. B **35**, 541 (1971).
84. — Ref. [59].
85. For a more extensive review than we can give here, compare Refs. [86, 87].
86. Sakurai, J. J.: Invited talk at the Japanese – U.S. – Joint Seminar, Stanford, February 1971.
87. — Erice Lectures, UCLA/71/TEP/39 (1971).
88. Jackson, Quigg: Phys. Letters **29** B, 236 (1969).
89. Bartl, A., Majerotto, W., Schildknecht, D.: DESY 72/4 submitted to Nucl. Phys.
90. Bingham, H. H., *et al.*: Phys. Rev. Letters **25**, 1233 (1970) and Ballam, J., *et al.*: SLAC-PUB-941 (1971) submitted to Phys. Rev.
91. Locher, M. P., Sandhas, W.: Z. Physik **195**, 461 (1966).
92. Stodolsky, L.: Phys. Rev. Letters **18**, 135 (1967), Freund, P. G. O.: Nuovo Cim. **44** A, 411 (1966).
93. Joos, H.: Phys. Letters **24** B, 103 (1967). – Kajantie, K., Trefil, J. S.: Phys. Letters **24** B, 106 (1967).
94. Alvensleben, H., *et al.*: Phys. Rev. Letters **25**, 1377 (1970); Nucl Phys. B **25**, 342 (1971).
95. Wolf, G.: Proc. of the Internat. Symposium on Electron and Photon Interactions at High Energies. Cornell University, Ithaca, N. Y. (1971).
96. ABBHHM Collaboration, Phys. Rev. **175**, 1669 (1968) and Phys. Rev. **188**, 2060 (1969).
97. Bulos, F., *et al.*: See data quoted by D. W. G. S. Leith in lectures presented to the Scottish Universities Summer School in Physics (1970).

98. SLAC-Berkeley-Tufts Collaboration, Phys. Rev. Letters **24**, 955 (1970); **24**, 960 (1970) and contributions to the International Symposium on Electron and Photon Interactions at High Energies. Cornell University, Ithaca, N.Y. (1971).
99. McClellan, G., et al.: Phys. Rev. Letters **22**, 374 (1969) and preprints CLNS-154 (1971) and CLNS-168 (1971).
100. Alvensleben, H., et al.: Phys. Rev. Letters **23**, 1058 (1969).
101. SLAC-Weizmann-Tel Aviv Coll.: Contribution to the International Symposium on Electron and Photon Interactions at High Energies, Cornell University Ithaca, N.Y. (1971).
102. Levi, E. M., Frankfurt, L. L.: JETP Letters **2** (1965). – Lipkin, H. J., Scheck, F.: Phys. Rev. Letters **16** (1965).
103. Buschhorn, G., et al.: DESY 71/51 (1971).
104. For theoretical work on ϱ^0 electroproduction in the framework of vector meson dominance compare Ref. [8a] and also Cho, C. F., Gounaris, G. J.: Phys. Rev. **186**, 1619 (1969). – Dieterle, B. D.: SLAC-PUB-595 for the question of the ϱ^0 slope especially Refs. [109–111].
105. Driver, C., et al.: Nucl. Phys. B **38**, 1 (1972).
106. Andrews, D., et al.: Phys. Rev. Letters **26**, 864 (1971) and CLNS-169 (1971).
107. Bloom, E. D., et al.: Phys. Rev. Letters **28**, 516 (1972).
108. Foley, K. J., et al.: Phys. Rev. Letters **11**, 425 (1963).
109. Cheng, H., Wu, T. T.: Phys. Rev. **183**, 1324 (1969).
110. Bjorken, J. D.: Proc. of the International Symposium on Electron and Photon Interactions at High Energies. Cornell University, Ithaca, N.Y. (1971).
111. For a critical recent discussion on the problem of change or no change of slope with k^2 in ϱ^0 electroproduction, see Nieh, H. T.: CERN Ref. TH 1432.
112. MIT-SLAC Report XV International Conference on High Energy Physics, Kiev USSR (1970) and review by Kendall, H. W.: International Symposium on Electron and Photon Interactions at High Energies. Cornell University, Ithaca, N.Y. (1971).
113. Feynman, R. P.: Unpublished. Bjorken, J. D., Paschos, E. A.: Phys. Rev. **185**, 1975 (1969). See also Gilman, F. J.: Proc. of the International Symposium on Electron and Photon Interactions at High Energies. Liverpool: 1969.
114. e.g. Wilson, K.: Proc. of the International Symposium on Electron and Photon Interactions at High Energies. Cornell University, Ithaca, N.Y. (1971).
115. For a different attempt to supplement vector meson dominance with light cone contributions within the framework of mass dispersion relations see the review given by Brandt, R., Preparata, G.: Lectures given at the Hamburg Summer Institute (1971).
116. See Refs. [7, 8].
117. See e.g. Gilman, F. J.: Ref. [113] for the definiton of νW_2.
118. Schwitters, R. F., et al.: Phys. Rev. Lett. **27**, 120 (1971).
119. Gottfried, K.: Proc. of the International Symposium on Electron and Photon Interactions at High Energies, Cornell University, Ithaca, N.Y. (1971).
120. Sakurai, J. J., Schildknecht, D.: UCLA/72/TEP/49 and to be published. For related work on higher mass vector state contributions compare Ref. [121].
121. Gribov, V. N.: JETP **30**, 709 (1970); Ritson, D. M.: Phys. Rev. D **3**, 1267 (1971); Fujikawa, K.: Phys. Rev. D **9**, 2794 (1971); Bjorken, J. D.: Ref. [110]; Brodsky, S. J., Purplin, J.: Phys. Rev. **182**, 1794 (1969).

Dr. D. Schildknecht
Deutsches Elektronen Synchrotron, DESY
D-2000 Hamburg 52

$\varrho - \omega$ Mixing

F. M. RENARD

Contents

I. Introduction

Mixing of elementary particles is an old problem. It occurs everytime when a symmetry is broken. Well known cases are the ones of the $K^0 - \bar{K}^0$ system (weak interactions violating strangeness and then CP) or of the $\eta - X$ and $\omega - \varphi$ systems (medium strong interactions violating SU_3).

The possibility of a $\varrho - \omega$ mixing was first pointed out by Glashow [1] in 1961. Isospin invariance is here broken by the electromagnetic interactions. States with the same quantum numbers including I_z but with different values of I can now mix; this is also the case of $\pi^0 - \eta$ or $\Sigma^0 - \Lambda$... etc., but the $\varrho - \omega$ case is particular due to the near degeneracy of the ϱ and ω masses which increases the effects.

The usual view of this phenomenon is the following. With strong interactions only, Isospin would be conserved and the physical states would be the orthonormal ones of the (I, I_z) basis: $|\varrho^0\rangle \equiv |1, 0\rangle$ and $|\omega^0\rangle \equiv |0, 0\rangle$ ($\langle \varrho^0 | \omega^0 \rangle = 0$). In this basis the hamiltonian or S-matrix or vector-meson propagator would be diagonal. However the physical situation includes the electromagnetic interactions. In first order in $\alpha = e^2/4\pi = 1/137$ transitions with $|\Delta I| = 1$ or 2 may occur (G-parity is violated). This allows for instance the decays $\omega^0 \to 2\pi$ or $\varrho^0 \to 3\pi$, for which the branching ratios are expected to be of order $\alpha^2 \simeq 0.5 \times 10^{-4}$. But in addition to these transitions $\langle 2\pi | S | \omega^0 \rangle$, $\langle 3\pi | S | \varrho^0 \rangle$ and partly due to them, non zero matrix elements $\langle \omega^0 | S | \varrho^0 \rangle$ exist also in order α. The mass-matrix is no more diagonal in the isospin basis; the physical states are no more the isospin states but are the new eigenstates of this mass matrix. The observed vector mesons ϱ and ω are the linear combinations of the states $|\varrho^0\rangle$ and $|\omega^0\rangle$ which diagonalize this matrix. They also correspond to the poles of the complete vector meson propagator.

One may first notice that no selection rule remains to impose $\langle \varrho | \omega \rangle$ to be zero and in fact it is not; this non orthogonality will correspond to a linear transformation which is not unitary; it is not a rotation angle which describes the mixing but a complex parameter.

One may also ask for $\varrho - \varphi$ mixing; in fact the complete problem involves the three ϱ, ω and φ vector mesons. But it is easy to verify that in order α, the $\varrho - \omega$ and $\varrho - \varphi$ problems are disconnected once one has solved the usual $\omega - \varphi$ one. Then, there are two reasons to leave for the moment the $\varrho - \varphi$ mixing: first the enhancement factor $[m_\varrho - m_\varphi]^{-1}$ is less than $[m_\varrho - m_\omega]^{-1}$ even by including the width; secondly the electromagnetic transition $\langle \varphi^0 | S | \varrho^0 \rangle$ is expected to be much weaker than the $\langle \omega^0 | S | \varrho^0 \rangle$ one on the basis of the decoupling of the φ from the states formed with non strange quarks.

The mixing formalism appropriate for overlapping resonances will be reviewed in Section II. It will be shown that up to small effects due to some energy dependances, the essential feature is given by the mixing parameter

$$\varepsilon = \frac{(m_\varrho + m_\omega)\, \delta m}{m_\omega^2 - m_\varrho^2 - i(m_\omega \Gamma_\omega - m_\varrho \Gamma_\varrho)}.$$

This is essentially the mass-mixing method but from a phenomenological point of view it describes also the main results of the general case including current-mixing (see Appendix A).

The quantity δm is the non diagonal term $\langle \omega^0 | S | \varrho^0 \rangle$. Its imaginary part is through the unitary relation related to the physical states into which ω^0 and ϱ^0 can decay. Its real part can only be evaluated with some model for the electromagnetic self-masses; it can also be related through higher symmetries to the meson mass differences (for example the Coleman-Glashow tadpole model) (see Section III).

The coupling constants of the physical ϱ and ω can now be completely expressed in terms of the ones of the isospin states. There will be generally two contributions; the "direct transition" from the primitive isospin state and the mixing term. The $\omega \to 2\pi$ and $\varrho \to 3\pi$ cases are particular as the direct transition is of purely electromagnetic origin and expected to be of order α; the mixing term proportionnal to $\varrho^0 \to 2\pi$ and $\omega^0 \to 3\pi$ respectively are expected (from the fact that $|\varepsilon| \gg \alpha$) to dominate. In addition for $\omega \to 2\pi$ the direct transition is cancelled by a corresponding contribution to δm. These features allow us (see Section IV) to predict in magnitude and in phase the amplitudes for the main decay modes of the ϱ and ω mesons. In particular non negligible phases appear for the $V - \gamma$ or $V - \pi - \gamma$ couplings.

The first comparison with experiment concerns the width $\Gamma_{\omega \to 2\pi}$. The mixing models with δm related to the electromagnetic mass difference predict a ratio $R = \Gamma_{\omega \to \pi^+ \pi^-} / \Gamma_{\omega \to 3\pi}$ of a few percent. During the period 1961–1968 much experimental effort had been done but the energy resolution and the poor statistics made the results much fluctuating. One had to wait until 1969 and the Orsay storage ring experiment $e^+ e^- \to 2\pi$ as well as the Berkeley experiment $\pi^+ p \to \pi^+ \pi^- + \varDelta^{++}$ to see this ratio confirmed. Many other groups have now seen this effect in various reactions, with $R \geq 1\%$.

The 2nd step concerns the phase predictions. In principle $e^+ e^- \to 2\pi$ or any other reaction in which the production amplitudes of vector mesons are known in relative magnitude and phase is adequate. The complete amplitude going through vector mesons may always be written as the sum of two Breit-Wigner formulas for the physical ϱ and ω mesons. The $\varrho - \omega$ interference effect is therefore completely determined by the phases of the production and of the decay amplitudes. In the case of $e^+ e^- \to$ hadrons (see Section V), the production amplitude is completely known in terms of the $V - \gamma$ couplings (the small phases are expressed in terms of ε); the decay amplitudes ($V \to 2\pi, 3\pi, \pi\gamma \ldots$) are also simply expressed in terms of ε. These reactions are in principle the best ones for testing the mixing formalism and for determining ε with accuracy. In fact the last Orsay measurement of $e^+ e^- \to \pi^+ \pi^-$ is also in agreement with the phase prediction for ε.

In the case of production of vector mesons (see Section VI and VII) the ideal program would be to analyze first the lepton pair production $A + B \to C + V_{\to e^+ e^-}$; the $\varrho - \omega$ interference term determines the relative phase of the ϱ and ω production amplitudes ($A + B \to C + \varrho$ and $A + B \to C + \omega$). Then the hadronic decay modes would be predicted; for example $A + B \to C + (\pi^+ + \pi^-)$ would be predicted in terms of ε and of these production amplitudes and the comparison with experiment would provide a subsequent check. The photoproduction of vector mesons on nuclei has already been exploited in this way, however two different experiments seem to give incompatible results, and new measurements are needed.

The last step would consist in checking more precisely the role of the "electromagnetic direct transition" or the energy dependance of ε or the possibility of a $\varrho - \varphi$ mixing. This needs an experimental accurary 10 times greater than the present ones, but the mixing mechanism would then be completely understood (see Section VIII).

The energy dependence of $W(s)$ necessitates approximations. Colemε and Schnitzer [7] have discussed two of them:

– the mass-mixing (or particle-mixing) which corresponds to neglect completely the s-dependence of W: $F(s) \simeq (s - W)^{-1}$; this is equivalent to introducing renormalisation terms (in particular non diagonal terms) in the massive part of the vector meson lagrangian.

– the current-mixing (or vector-mixing or kinetic-mixing) which corresponds to a linear s-dependence: $W \simeq m_0^2 + s\delta$, $F(s) \simeq [s(1 - \delta) - m_0^2]^{-1}$; this is equivalent to renormalizing the kinetic part of the vector meson lagrangian.

In the present $\varrho - \omega$ case, the mass-mixing approach has been first considered [8, 9]; but it has been shown [10] that for a purely phenomenological treatment and up to small terms like $\Gamma_\varrho/2m_\varrho$, the results of both approximations are the same. This is due to the fact that one has to deal with $\varrho - \omega$ interferences which are located in a small energy range (Γ_ω, or $m_\omega - m_\varrho$) in which the energy dependence of $W(s)$ is negligible. In the following we shall therefore neglect it; this is equivalent formally to the use of the mass-mixing method [9]. The formulas corresponding to the general case (mass + kinetic mixing) will be given in Appendix A.

In the isospin basis $|j\rangle$ ($|\varrho^0\rangle \equiv |1, 0\rangle$ and $|\omega^0\rangle \equiv |0, 0\rangle$) the propagator and the mass-matrix W would be diagonal if there were no electro magnetic interactions (E.M.I.). In the presence of E.M.I. they are not diagonal but still symmetric if time reversal invariance holds (this will always be supposed in the following but the general case without time reversal invariance has also been considered [11]).

The physical states are the eigenstates $|a\rangle$ which are the linear combinations of $|j\rangle$ that diagonalize W:

$$|a\rangle = \sum_j C_{aj} |j\rangle \qquad W|a\rangle = z_a |a\rangle$$

C is a 2×2 complex matrix which represents these linear combinations. C is generally not a unitary matrix ($C^+ \neq C^{-1}$), then the left eigenvectors are not the conjugate of the right ones:

$$\langle \tilde{a}| W = z_a \langle \tilde{a}| \qquad \text{with} \qquad \langle \tilde{a}| = \sum_j (C^{-1})_{ja} \langle j|.$$

But as W is aymmetric the $\langle \tilde{a}|$ are simply the transposed (in the isospin basis) of the $|a\rangle$.

One can normalize the states such that

$$\langle \tilde{a}|b\rangle = \delta_{ab} \qquad \text{and} \qquad I = \sum_a |\tilde{a}\rangle \langle a| = \sum_a |a\rangle \langle \tilde{a}|$$

and then

$$F(s) = (s - W)^{-1} = \sum_a |a\rangle \langle \tilde{a}|/(s - z_a).$$

II. The Mixing Formalism

Three methods have been used to treat this problem. The fi:
based on the time dependent Wigner-Weisskopf method [2] w
coupled Schrödinger equations. It is more convenient for th
instable particles for which one can observe the evolution in
example the $K\bar{K}$ problem, but it has been also used for the $\varrho - c$
by Feinberg and Bernstein [3] and more recently by Horn [4

In the case where one measures functions of energy like cro:
showing strong interaction resonances, a formalism givin
the scattering amplitudes is preferred. In the non-relativistic f
this is the method proposed by Stodolsky [5] for treating the μ
overlapping resonances. The relativistic fromulation is the p
method [6].

Both of these last two methods study the properties of a m
$W(s)$. In the propagator formalism it occurs as follows. Co
vector meson propagator:

$$\Delta_{\mu\nu}(q) = \int d_4 x \, e^{-iq \cdot x} \langle 0| T(\phi_\mu(x) \, \phi_\nu(0))|0\rangle ,$$

it satisfies the spectral representation:

$$\Delta_{\mu\nu}(q) = \int\limits_{s_0}^{\infty} ds' \, \frac{\varrho(s')}{q^2 - s'} (g_{\mu\nu} - q_\mu q_\nu/s') \equiv F(s) \, g_{\mu\nu} + s^{-1}(F(0) -)$$

where $s \equiv q^2$.

The interesting propagator function $F(s)$ may then t
$F(s) = [s - W(s)]^{-1}$; $W(s) = m_0^2 + \delta m^2$ is then the sum of
meson bare mass and of all the self-energies corresponding to
diagrams of Fig. 1. δm^2 is a complex quantity whose imag

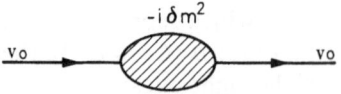

Fig. 1. Contribution of self-energies to the one-particle S-matri:

is related to the real (physical) intermediate states occur
diagram. The physical vector meson state correspond to 1
$F(s)$ (located in the 2nd sheet of the complex energy plane, in
resonances).

In the many channel case $F(s)$ and $W(s)$ are matrices. T
states correspond to zeros of det$[s - W(s)]$; they are then the e
of the mass-matrix $W(s)$.

The energy dependence of $W(s)$ necessitates approximations. Coleman and Schnitzer [7] have discussed two of them:

– the mass-mixing (or particle-mixing) which corresponds to neglect completely the s-dependence of W: $F(s) \simeq (s - W)^{-1}$; this is equivalent to introducing renormalisation terms (in particular non diagonal terms) in the massive part of the vector meson lagrangian.

– the current-mixing (or vector-mixing or kinetic-mixing) which corresponds to a linear s-dependence: $W \simeq m_0^2 + s\delta$, $F(s) \simeq [s(1 - \delta) - m_0^2]^{-1}$; this is equivalent to renormalizing the kinetic part of the vector meson lagrangian.

In the present $\varrho - \omega$ case, the mass-mixing approach has been first considered [8, 9]; but it has been shown [10] that for a purely phenomenological treatment and up to small terms like $\Gamma_\varrho/2m_\varrho$, the results of both approximations are the same. This is due to the fact that one has to deal with $\varrho - \omega$ interferences which are located in a small energy range (Γ_ω, or $m_\omega - m_\varrho$) in which the energy dependence of $W(s)$ is negligible. In the following we shall therefore neglect it; this is equivalent formally to the use of the mass-mixing method [9]. The formulas corresponding to the general case (mass + kinetic mixing) will be given in Appendix A.

In the isospin basis $|j\rangle$ ($|\varrho^0\rangle \equiv |1, 0\rangle$ and $|\omega^0\rangle \equiv |0, 0\rangle$) the propagator and the mass-matrix W would be diagonal if there were no electro magnetic interactions (E.M.I.). In the presence of E.M.I. they are not diagonal but still symmetric if time reversal invariance holds (this will always be supposed in the following but the general case without time reversal invariance has also been considered [11]).

The physical states are the eigenstates $|a\rangle$ which are the linear combinations of $|j\rangle$ that diagonalize W:

$$|a\rangle = \sum_j C_{aj} |j\rangle \qquad W|a\rangle = z_a |a\rangle$$

C is a 2×2 complex matrix which represents these linear combinations. C is generally not a unitary matrix ($C^+ \neq C^{-1}$), then the left eigenvectors are not the conjugate of the right ones:

$$\langle \tilde{a}| W = z_a \langle \tilde{a}| \quad \text{with} \quad \langle \tilde{a}| = \sum_j (C^{-1})_{ja} \langle j| .$$

But as W is aymmetric the $\langle \tilde{a}|$ are simply the transposed (in the isospin basis) of the $|a\rangle$.

One can normalize the states such that

$$\langle \tilde{a}|b\rangle = \delta_{ab} \quad \text{and} \quad I = \sum_a |\tilde{a}\rangle \langle a| = \sum_a |a\rangle \langle \tilde{a}|$$

and then

$$F(s) = (s - W)^{-1} = \sum_a |a\rangle \langle \tilde{a}|/(s - z_a) .$$

II. The Mixing Formalism

Three methods have been used to treat this problem. The first one is based on the time dependent Wigner-Weisskopf method [2] which uses coupled Schrödinger equations. It is more convenient for the case of instable particles for which one can observe the evolution in time, for example the $K\bar{K}$ problem, but it has been also used for the $\varrho - \omega$ problem by Feinberg and Bernstein [3] and more recently by Horn [4].

In the case where one measures functions of energy like cross sections showing strong interaction resonances, a formalism giving directly the scattering amplitudes is preferred. In the non-relativistic framework this is the method proposed by Stodolsky [5] for treating the problem of overlapping resonances. The relativistic fromulation is the propagator method [6].

Both of these last two methods study the properties of a mass matrix $W(s)$. In the propagator formalism it occurs as follows. Consider the vector meson propagator:

$$\Delta_{\mu\nu}(q) = \int d_4 x \, e^{-iq \cdot x} \langle 0| T(\phi_\mu(x) \, \phi_\nu(0))|0\rangle ,$$

it satisfies the spectral representation:

$$\Delta_{\mu\nu}(q) = \int_{s_0}^{\infty} ds' \, \frac{\varrho(s')}{q^2 - s'} (g_{\mu\nu} - q_\mu q_\nu/s') \equiv F(s) \, g_{\mu\nu} + s^{-1}(F(0) - F(s)) \, q_\mu q_\nu$$

where $s \equiv q^2$.

The interesting propagator function $F(s)$ may then be written: $F(s) = [s - W(s)]^{-1}$; $W(s) = m_0^2 + \delta m^2$ is then the sum of the vector meson bare mass and of all the self-energies corresponding to the proper diagrams of Fig. 1. δm^2 is a complex quantity whose imaginary part

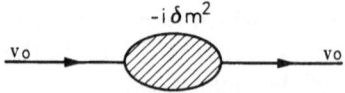

Fig. 1. Contribution of self-energies to the one-particle S-matrix

is related to the real (physical) intermediate states occuring in this diagram. The physical vector meson state correspond to the pole of $F(s)$ (located in the 2nd sheet of the complex energy plane, in the case of resonances).

In the many channel case $F(s)$ and $W(s)$ are matrices. The physical states correspond to zeros of det $[s - W(s)]$; they are then the eigenvectors of the mass-matrix $W(s)$.

Starting from the orthonormalized isospin basis $\langle \varrho^0 | \varrho^0 \rangle = \langle \omega^0 | \omega^0 \rangle = 1$, $\langle \varrho^0 | \omega^0 \rangle = 0$, one can completely reduce the arbitrary part of the matrix C by doing the convenient choice of phases:

$$
\begin{aligned}
|\varrho\rangle = |\varrho^0\rangle - \varepsilon |\omega^0\rangle \qquad \langle \tilde{\varrho}| = \langle \varrho^0| - \varepsilon \langle \omega^0| \\
|\omega\rangle = |\omega^0\rangle + \varepsilon |\varrho^0\rangle \qquad \langle \tilde{\omega}| = \langle \omega^0| + \varepsilon \langle \varrho^0|
\end{aligned}
\tag{1}
$$

(the normalisation factors $(1 + \varepsilon^2)^{-\frac{1}{2}}$ are omitted as they are 2nd order effect $< 10^{-3}$). ε is the complex non diagonal element of the C matrix:

$$
C = \begin{pmatrix} 1 - \varepsilon \\ \varepsilon & 1 \end{pmatrix} \qquad W = \begin{pmatrix} W_{\varrho^0 \varrho^0} & W_{\varrho^0 \omega^0} \\ W_{\varrho^0 \omega^0} & W_{\omega^0 \omega^0} \end{pmatrix} \qquad W_D = C W C^{-1} \equiv \begin{pmatrix} z_\varrho & 0 \\ 0 & z_\omega \end{pmatrix}.
$$

In first order, one has also the relation:

$$
\varepsilon = W_{\varrho^0 \omega^0} / (z_\omega - z_\varrho) .
$$

Notice, as announced in the introduction, that $\langle \omega | \varrho \rangle = -2i \, \mathrm{Im} \, \varepsilon \neq 0$. z_ϱ and z_ω, the positions of the poles, are the complex masses of the ϱ, ω physical states. From unitary and analyticity of $F(s)$, one relates of course their imaginary part to the observed width and recover the usual Breit-Wigner approximation:

$$
\begin{aligned}
-2i \, \mathrm{Im} \, W = F^{-1}(s + i\eta) \, 2i\pi\varrho(s) \, F^{-1}(s - i\eta) \\
= 2i\pi \sum_n \langle 0|j(0)|n\rangle \langle n|j(0)|0\rangle \equiv 2i \sqrt{s} \, \Gamma(s) ,
\end{aligned}
$$

and in the physical basis

$$
-\,\mathrm{Im} \, z_a \simeq m_a \Gamma_a
$$

then,

$$
z_\varrho \simeq m_\varrho^2 - i m_\varrho \Gamma_\varrho , \qquad z_\omega \simeq m_\omega^2 - i m_\omega \Gamma_\omega
$$

and

$$
\varepsilon = \frac{W_{\varrho^0 \omega^0}}{m_\omega^2 - m_\varrho^2 - i(m_\omega \Gamma_\omega - m_\varrho \Gamma_\varrho)} .
\tag{2}
$$

The transition amplitude with intermediate vector mesons will then be written:

$$
\langle f | F(s) | i \rangle = \sum_{a = \varrho, \omega} \frac{\langle f | a \rangle \langle \tilde{a} | i \rangle}{s - m_a^2 + i m_a \Gamma_a} .
\tag{3}
$$

This is the result already proved by Stodolsky [5] and which is valid for any superposition of resonances; the scattering matrix can always be written as the sum of Breit-Wigner amplitudes corresponding to the physical states.

III. Models for the Mixing Parameter

The only unknown quantity occuring in the expression of ε is the complex off-diagonal term $W_{\varrho^0\omega^0}$.

In the following, let us define $W_{\varrho^0\omega^0} \equiv (m_\varrho + m_\omega)(\delta m_R + i\delta m_I)$ $\simeq 2m_\varrho(\delta m_R + i\delta m_I)$. The imaginary part is however related [9] to the physical intermediate states coupled to the isospin states ϱ^0 and ω^0:

$$-2im_\varrho\delta m_I = \tfrac{1}{2}\langle\varrho^0|W^+ - W|\omega^0\rangle = i\pi\sum_n\langle\varrho^0|R^+|n\rangle\langle n|R|\omega^0\rangle.$$

The important intermediate states n are those which appears in Fig. 2 and which constitute the main decay modes of the vector mesons.

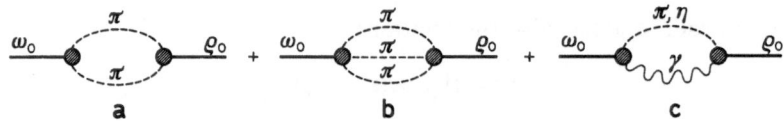

Fig. 2. Contributions to the imaginary part δm_I

In 2a and 2b appear the "direct transitions" $\omega^0 \to 2\pi$ and $\varrho^0 \to 3\pi$; they are normally due to E.M.I. independantly of the possibility of mixing. One expects for the coupling constants: $|f_{\omega^0\pi\pi}| \simeq \alpha|f_{\varrho^0\pi\pi}|$ and $|f_{\varrho^0 3\pi}| \simeq \alpha|f_{\omega^0 3\pi}|$, then the contributions of Fig. 2 can be written:

$$-2\delta m_I = \frac{f_{\omega^0\pi\pi}}{f_{\varrho^0\pi\pi}}\,\Gamma_{\varrho^0\to 2\pi} + \frac{f_{\varrho^0 3\pi}}{f_{\omega^0 3\pi}}\,\Gamma_{\omega^0\to 3\pi} + \sqrt{\Gamma_{\omega^0\to\pi\gamma}\Gamma_{\varrho^0\to\pi\gamma}} + \cdots. \quad (4)$$

numerically:

$$-2\delta m_I \simeq \pm 0.8 \pm 0.1 + 0.3 \text{ MeV}$$

(if one uses the V.D.M. predictions [12] for the unknown $V \to P + \gamma$ decay rates). The signs of the first two couplings are unknown and uncorrelated, then one has reasonably the upper limit:

$$|\delta m_I| \lesssim 0.6 \text{ MeV}. \quad (5)$$

No limitation a priori exists for the real part δm_R; this is a self-mass due to virtual intermediate states similar to the ones responsible of the E.M. mass differences; one may use to predict δm_R the same models or symmetry relations as for the E.M. meson mass differences [13]. Here one is concerned with a $|\Delta I| = 1$ transition of the same kind as in $m_{K^+} - m_{K^0}$, $m_{K^{*+}} - m_{K^{*0}}$, or $m_p - m_n \dots$ etc.

Consider first the possible intermediate states between ϱ^0 and ω^0. The V.D.M. contribution has been calculated by many authors [14] and gives now (see Fig. 3):

$$2\delta m_R = e^2 f_{\varrho\gamma} f_{\omega\gamma}/m_\omega^3 \simeq 0.9 \text{ MeV} .$$

The other intermediate states are those which appear in the virtual Compton scattering on vector mesons (Fig. 4). It has been proposed by Coleman and Glashow [15] that the dominant part is due to the tadpole mechanism (Fig. 5). This tadpole diagram may effectively be due to the

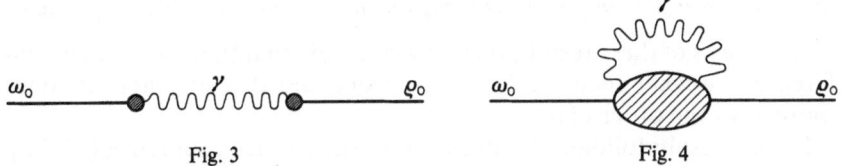

Fig. 3   Fig. 4

Fig. 3. Vector dominance contribution to the real part δm_R

Fig. 4. Compton contributions to δm_R

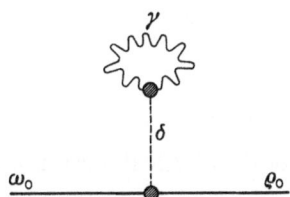

Fig. 5. Tadpole contribution to δm_R

exchange of the $\delta(975)\, 0^{++}, I = 1$ meson which appears as an $\eta\pi$ resonance. It can also be explained by a Regge pole exchange. The first calculation of Coleman and Glashow gives:

$$\delta m_R = -3^{-\frac{1}{2}}(m_{K*}^2 - m_\varrho^2)\, R/m_\varrho ,$$

with $R \simeq 0.02$ calculated from the baryon E.M. mass differences; this gives $\delta m_R \simeq -2.5$ MeV. A similar, but somewhat greater, order of magnitude comes if one uses the quark-model with additivity for the δ couplings and ideal $\omega - \varphi$ mixing (in the convention $\sin\theta_V = +3^{-\frac{1}{2}}$):

$$\delta m_R = m_{K*+} - m_{K*0} = -8 \pm 3 \text{ MeV} \ [16] .$$

An other approach starts from Unitary Symmetry with U-spin invariance of the E.M.I., and the relation [17].

$$\sqrt{3}\,(m_\varrho + m_\omega)\, \{\cos\theta_V\, \delta m_R^{\varrho\varphi} + \sin\theta_V\, \delta m_R^{\varrho\omega}\} = m_{\varrho^0}^2 - m_{\varrho^+}^2 + m_{K*+}^2 - m_{K*0}^2 . \quad (6)$$

The $\varrho\varphi$ mixing term can be neglected on the basis of the pure strange quark component of the φ; this relation involves however the $\varrho^0 - \varrho^+$ mass difference which is experimentally unknown. This is a $|\varDelta I| = 2$ transition in which the tadpole diagram does not occur. A sum rule [18] gives:

$$m_{\varrho^+} - m_{\varrho^0} = 1.5 \text{ MeV},$$

and a quark model calculation [19] gives the relation:

$$m_{\varrho^+} - m_{\varrho^0} = m_{\pi^+} - m_{\pi^0} + \tfrac{3}{2}(m_{K^0} - m_{K^+} + m_{K^{*+}} - m_{K^{*0}}) \simeq -1.5 \pm 4.5 \text{ MeV}.$$

On the basis of these results, one may conclude that the $\varrho^0 - \varrho^+$ mass difference is weaker than the $K^{*+} - K^{*0}$ one and that δm_R appears once more as of the order of $m_{K^{*+}} - m_{K^{*0}}$.

Similar results follow also from asymptotic symmetry sum rules [13, 20]. One concludes this section in noting that $|\delta m_R| \gg |\delta m_I|$; $W_{\varrho^0\omega^0}$ is then essentially real and negative; δm_R is of the order of a few Mev. This predicts definitely the magnitude (a few percent) and the phase (the one of $[z_\varrho - z_\omega]^{-1}$) of ε.

IV. Vector Meson Decays

The coupling constants of the physical mesons can be related to the ones of the isospin states:

$$\langle F|\varrho\rangle = \langle F|\varrho^0\rangle - \varepsilon\langle F|\omega^0\rangle \quad \text{and} \quad \langle F|\omega\rangle = \langle F|\omega^0\rangle + \varepsilon\langle F|\varrho^0\rangle.$$

Every coupling is then related to both ones of the isospin states:
  — the "direct transition": $\langle F|\varrho^0\rangle$ for the ϱ or $\langle F|\omega^0\rangle$ for the ω.
  — the transition due to mixing: $-\varepsilon\langle F|\omega^0\rangle$ for the ϱ or $\varepsilon\langle F|\varrho^0\rangle$ for the ω.

The presence of the complex mixing parameter ε leads to a non zero phase for the ϱF and ωF couplings, even if one starts with real couplings for the isospin states.

a) $\varrho, \omega \to 2\pi$: $f_{\omega\pi\pi} = f_{\omega^0\pi\pi} + \varepsilon f_{\varrho^0\pi\pi}$ $f_{\varrho\pi\pi} = f_{\varrho^0\pi\pi}$ in first order in α.

The direct transition $f_{\omega^0\pi\pi}$ is an unknown quantity, expected to be of order $\alpha f_{\varrho^0\pi\pi}$. However in $f_{\omega\pi\pi}$ it appears also in the imaginary part of $W_{\varrho^0\omega^0}$ which contributes to ε and cancels up to $\Gamma_\omega/\Gamma_\varrho$ or $(m_\omega - m_\varrho)/\Gamma_\varrho$ factors due to the fact that ω and ϱ are quasidegenerate and that the 2π channel dominates the ϱ decay [9]:

$$f_{\omega\pi\pi} \simeq f_{\omega^0\pi\pi} - \frac{2m_\varrho(\delta m_R + i\delta m_I)}{2m_\varrho(m_\varrho - m_\omega) - im_\varrho(\Gamma_\varrho - \Gamma_\omega)} f_{\varrho^0\pi\pi} \simeq f_{\omega^0\pi\pi}$$
$$- i\Gamma_\varrho^{-1}[2\delta m_R - i(f_{\omega^0\pi\pi}\Gamma_{\varrho^0 \to 2\pi}/f_{\varrho^0\pi\pi} + \cdots)] f_{\varrho^0\pi\pi}$$

then

$$f_{\omega\pi\pi} \simeq \varepsilon' f_{\varrho^0\pi\pi} \quad \text{and} \quad \Gamma_{\omega \to 2\pi} \simeq |\varepsilon'|^2 \, \Gamma_{\varrho \to 2\pi} \,.$$

ε' is ε without the $f_{\omega^0\pi\pi}$ contribution in δm_I; it is therefore the best known quantity. In any case as one expects $|\delta m_R| \gg |\delta m_I|$, one expects also $\varepsilon' \simeq \varepsilon$.

The experimental results [31], for $\Gamma_{\omega \to 2\pi}$ only, confirm these predictions. $\Gamma_{\omega \to 2\pi} \simeq 0.4$ MeV means $|\varepsilon'| \simeq 0.06$ and $|\delta m_R + i\delta m_I'| \simeq 3.6$ MeV. But as $|\delta m_I'| < 0.2$ MeV (the part which does not include the 2π intermediate state), one gets therefore from experiment

$$|\delta m_R| \simeq 3.6 \text{ MeV}$$

in agreement with the models (which predict in addition a negative sign). On this purely phenomenological ground, one deduces the phase of ε and ε'. $\delta m_R = -3.6$ MeV, $|\delta m_I'| < 0.2$ MeV and $|\delta m_I| < 0.6$ MeV and the usual masses and width of the ϱ and ω mesons give:

$$98° < \varphi_{\varepsilon'} < 104° \quad \text{and} \quad 91° < \varphi_{\varepsilon} < 111° \,. \tag{7}$$

These limits are only due to the upper limits on the imaginary parts. One may slightly enlarge (but not much) them by including experimental errors on the masses and width of the vector mesons and on $\Gamma_{\omega \to 2\pi}$. (With the opposite sign for δm_R, one would get $\varphi_{\varepsilon}' \simeq -79 \pm 3°$ and $\varphi_{\varepsilon} \simeq -79 \pm 10°$).

b) $\varrho, \omega \to 3\pi$: $f_{\varrho, 3\pi} = f_{\varrho^0, 3\pi} - \varepsilon f_{\omega^0, 3\pi} \quad f_{\omega, 3\pi} = f_{\omega^0, 3\pi}.$

The "direct transition" $f_{\varrho^0, 3\pi}$ does not cancel in this case, as the 3π channel (and partial width) does not dominate. However from $|f_{\varrho^0, 3\pi}| \simeq \alpha |f_{\omega^0, 3\pi}|$ and $|\varepsilon| \gg \alpha$, it is expected (say within 12% error) that:

$$f_{\varrho, 3\pi} \simeq -\varepsilon f_{\omega, 3\pi}$$

and $\Gamma_{\varrho \to 3\pi} \simeq |\varepsilon|^2 \, \Gamma_{\omega \to 3\pi} \simeq 0.04$ MeV.

With the 2π and 3π decay modes one can test this dominance of the mixing term over the direct transitions by forming the following quantity:

$$f_{\omega\pi\pi}'/f_{\varrho\pi\pi} + f_{\varrho, 3\pi}/f_{\omega, 3\pi} = f_{\omega^0\pi\pi}/f_{\varrho^0\pi\pi} + f_{\varrho^0, 3\pi}/f_{\omega^0, 3\pi} \tag{8}$$

which should be only of order α, as the influence of ε has cancelled.

c) Electromagnetic Decays $\varrho, \omega \to \pi\gamma, \eta\gamma, e^+ e^- \ldots$

$f_{\varrho F} = f_{\varrho^0 F} - \varepsilon f_{\omega^0 F}$ and $f_{\omega F} = f_{\omega^0 F} + \varepsilon f_{\varrho^0 F}$; for any final state F of this kind both transitions $f_{\varrho^0 F}$ and $f_{\omega^0 F}$ occur in the same order. Nevertheless two interesting features occur due to the presence of ε. First some deviations from any prediction made for $f_{\varrho^0 F}$ and $f_{\omega^0 F}$ will appear when measuring $f_{\varrho F}$ and $f_{\omega F}$. For example if $f_{\varrho^0 \pi\gamma} \ll f_{\omega^0 \pi\gamma}$, then $f_{\varrho\pi\gamma} = f_{\varrho^0\pi\gamma} - \varepsilon f_{\omega^0\pi\gamma}$ will get a non negligible contribution from $\omega^0 \to \pi\gamma$. The situation is the conversed for the e^+e^- mode as $f_{\omega^0\gamma} < f_{\varrho^0\gamma}$. The SU_3 predictions, which are made for ω^0 and ϱ^0, are not exactly satisfied by ω and ϱ. Secondly non negligible phases appear due to the phase of ε. Starting for example from real $f_{\varrho^0\gamma}$ and $f_{\omega^0\gamma}$ couplings, with $\varphi_\varepsilon \simeq 101°$, one gets

$$\varphi_{\varrho\gamma} \equiv \text{phase} (f_{\varrho\gamma}) \simeq -1° \quad \text{and} \quad \varphi_{\omega\gamma} \equiv \text{phase} (f_{\omega\gamma}) \simeq +10°. \quad (9)$$

These phases will not be negligible in lepton pairs photoproduction.

V. $e^+ e^-$ Reactions

a) $e^+ e^- \to 2\pi$

In the vicinity of the ϱ, ω masses the important diagrams are those of Fig. 6. From the general form (3) of any amplitude, one gets in this case:

$$R_{fi} = \frac{f_{\varrho\pi\pi} f_{\tilde{\varrho}\gamma}}{s - m_\varrho^2 + im_\varrho \Gamma_\varrho} + \frac{f_{\omega\pi\pi} f_{\tilde{\omega}\gamma}}{s - m_\omega^2 + im_\omega \Gamma_\omega} \quad (10)$$

$$\equiv f_{\varrho\pi\pi} f_{\tilde{\varrho}\gamma} [(s - m_\varrho^2 + im_\varrho \Gamma_\varrho)^{-1} + \xi e^{i\phi} (s - m_\omega^2 + im_\omega \Gamma_\omega)^{-1}].$$

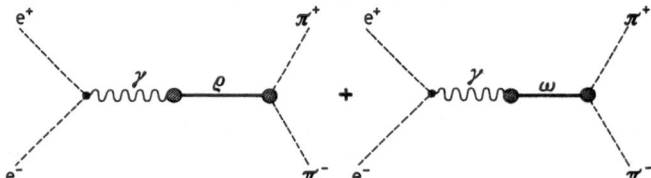

Fig. 6. Diagrams for $e^+ e^- \to 2\pi$ in the vicinity of the ϱ, ω masses

The quantity $\xi e^{i\phi} \equiv (f_{\tilde{\omega}\gamma}/f_{\tilde{\varrho}\gamma})(f_{\omega\pi\pi}/f_{\varrho\pi\pi})$ governs the shape of the interference and then of the cross-section around the ω mass. (see Fig. 7). Knowing $|f_{\tilde{\omega}\gamma}|, |f_{\tilde{\varrho}\gamma}|, \Gamma_\varrho$ and Γ_ω from the whole $e^+ e^- \to 2\pi$ and 3π productions [21], the Orsay storage ring group is able to relate ξ with $|\varepsilon'|$ and gives the result [22]:

$$R = \Gamma_{\omega \to 2\pi}/\Gamma_{\omega \to 3\pi} = 3.3^{+2.1}_{-1.6} \% \quad \text{this is equivalent to } |\varepsilon'| = 0.06 \pm 0.015$$

which has been used in Section IV.

They also give

$$\phi = 89° \pm 15°,$$

which, using $\phi \equiv \varphi_{\omega\gamma} - \varphi_{\varrho\gamma} + \varphi_{\varepsilon'}$ should be compared with the predictions (7) and (9): $\phi = 10° + 1° + 101° = 112°$, in rather good agreement.

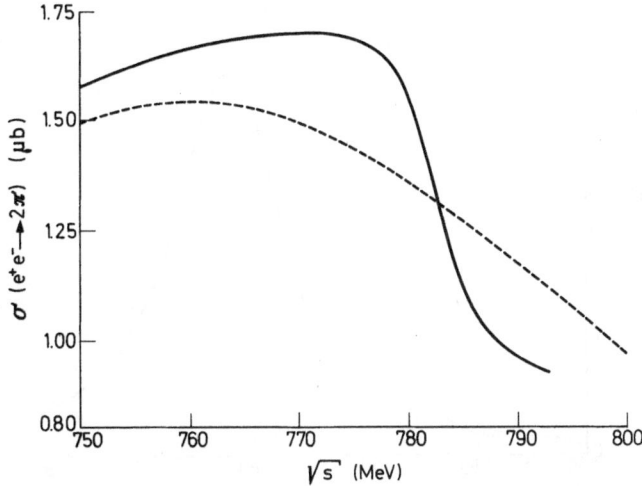

Fig. 7. Interference shape in $e^+ e^- \to 2\pi$

---- ϱ term only

—— $\varrho + \omega$ terms

b) $e^+ e^- \to 3\pi$

The amplitude is now

$$R_{fi} = \frac{f_{\varrho, 3\pi} f_{\tilde{\varrho}\gamma}}{s - m_\varrho^2 + i m_\varrho \Gamma_\varrho} + \frac{f_{\omega, 3\pi} f_{\tilde{\omega}\gamma}}{s - m_\omega^2 + i m_\omega \Gamma_\omega} \tag{11}$$

$$\equiv f_{\omega, 3\pi} f_{\tilde{\omega}\gamma} [(s - m_\omega^2 + i m_\omega \Gamma_\omega)^{-1} + \xi e^{i\phi} (s - m_\varrho^2 + i m_\varrho \Gamma_\varrho)^{-1}] \, .$$

The factor $\xi e^{i\phi} \equiv (f_{\tilde{\varrho}\gamma}/f_{\tilde{\omega}\gamma}) (f_{\varrho, 3\pi}/f_{\omega, 3\pi})$ is now expected with:

$$\xi \simeq |f_{\tilde{\varrho}\gamma}/f_{\tilde{\omega}\gamma}| \, |\varepsilon| \simeq 2.65 \, |\varepsilon| \simeq 0.16 \quad \text{and} \quad \phi = \varphi_{\varrho\gamma} - \varphi_{\omega\gamma} + \varphi_\varepsilon - \pi = -90°$$

The kind of effect is here completely different as the 2nd term of R_{fi} gives a broad back-ground under the narrow ω peak. This leads essentially to a small renormalization of the ω amplitude (see Fig. 8). For example the factor:

$$r \equiv 1 + \xi e^{i\phi} \frac{s - m_\omega^2 + i m_\omega \Gamma_\omega}{s - m_\varrho^2 + i m_\varrho \Gamma_\varrho}$$

is $1 + 0.002 - i\, 0.0015$ at $s = m_\omega^2$ and $1 + 0.035 - i\, 0.015$ at $s = m_\varrho^2$.

c) Other processes $e^+ e^- \to \pi\gamma, \eta\gamma \ldots$

The modifications occur essentially through the ones of the vector meson coupling constants $V - \gamma$ and $V - P - \gamma$ (in modulus and in phase) as described in Section IV. At a 10% accuracy they are not negligible.

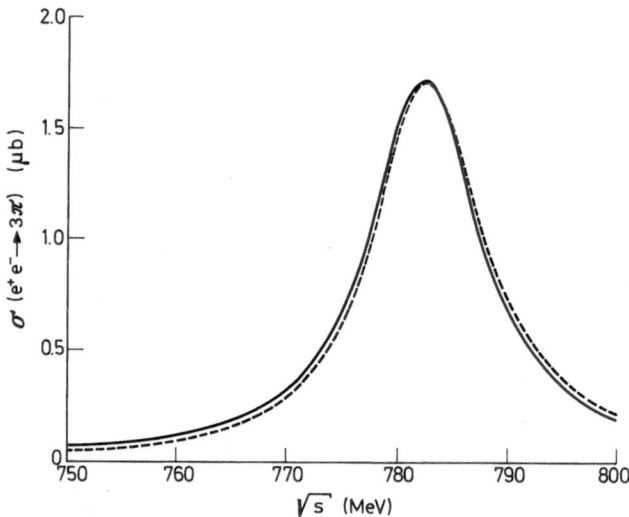

Fig. 8. Interference shape in $e^+e^- \to 3\pi$
$$---- \omega \text{ term only}$$
$$\underline{\quad\quad} \omega + \varrho \text{ terms}$$

For concluding this section, let us say that from a theoretical point of view the e^+e^- reactions are surely the best ones for determining the mixing parameter; ε is the only unknown quantity, no phase (except those ones due to ε itself) enters in the production amplitude.

VI. Vector Meson Photoproduction

This is a privileged reaction as the phases of the production amplitudes can be precisely determined. Consider the reaction $\gamma + B \to C + V_{\to F}$; in the vicinity of the ϱ and ω masses for $s \equiv p_V^2$, the amplitude has the general form (3):

$$R_{fi} = \frac{\langle F|\varrho\rangle\langle\tilde{\varrho}C|B\gamma\rangle}{s - m_\varrho^2 + im_\varrho\Gamma_\varrho} + \frac{\langle F|\omega\rangle\langle\tilde{\omega}C|B\gamma\rangle}{s - m_\omega^2 + im_\omega\Gamma_\omega}u + R_B \qquad (12)$$

in which a background amplitude R_B has been introduced; for example in the case of lepton pairs production $(F \equiv e^+e^-)$ R_B will be the Bethe-Heitler amplitude.

In this discussion we shall neglect the helicity dependence of the matrix elements. It may be important to consider it in particular for describing the degree of coherence c of the ϱ and ω amplitudes when they do not correspond to the same production mechanism. In this case a density matrix analysis [23] is required. The essential formulas are given in Appendix B, but along this section one will suppose $c = 1$ for simplicity.

The relative phase of the two terms of (12) is:

$$\phi_F \equiv \mathrm{Arg} \left[\frac{\langle F|\omega\rangle \langle \tilde{\omega}C|B\gamma\rangle}{\langle F|\varrho\rangle \langle \tilde{\varrho}C|B\gamma\rangle} \right] \equiv \phi_P + \phi_D \qquad (13)$$

where $\phi_D \equiv \mathrm{Arg}\,(\langle F|\omega\rangle/\langle F|\varrho\rangle)$ and $\phi_P \equiv \mathrm{Arg}\,(\langle \tilde{\omega}C|B\gamma\rangle/\langle \tilde{\varrho}C|B\gamma\rangle)$ $\equiv \phi_P^\omega - \phi_P^\varrho$ are the relative phase at the decay and at the production.

The decay phase is expressed in terms of the mixing parameter; for example:

$$\text{for } F \equiv e^+e^- \quad \phi_D = \mathrm{Arg}\,(f_{\omega\gamma}/f_{\varrho\gamma}) = \varphi_{\omega\gamma} - \varphi_{\varrho\gamma} \simeq +11°, \qquad (14)$$

$$\text{for } F \equiv \pi^+\pi^- \quad \phi_D = \mathrm{Arg}\,(f_{\omega\pi\pi}/f_{\varrho\pi\pi}) = \varphi_{\varepsilon'} \simeq 101° \pm 3°. \qquad (15)$$

The production phases depend upon the production mechanism as well as upon mixing. The measurement of the $\varrho - \omega$ interference gives $\phi_F = \phi_P + \phi_D$. With two decay modes ($F = e^+e^-$ and $F = \pi^+\pi^-$) one can eliminate ϕ_P and check the phase of ε:

$$\phi_{e^+e^-} - \phi_{\pi^+\pi^-} = \varphi_{\omega\gamma} - \varphi_{\varrho\gamma} - \varphi_{\varepsilon'}. \qquad (16)$$

The difficult task is now the comparison with experiment [31].

For lepton pairs photoproduction:

A Daresbury experiment [24, 25] on Carbon gives

$$\phi_{e^+e^-} = 100°{}^{+38°}_{-30°} \quad \text{at} \quad 3.5 \text{ Gev} \quad \text{and} \quad 118°{}^{+13°}_{-22°} \quad \text{at} \quad 4 \text{ Gev}.$$

A Desy-MIT experiment [26] on Beryllium gives

$$\phi_{e^+e^-} = 41° \pm 20° \quad \text{at} \quad 5.1 \text{ Gev}.$$

For pion pairs photoproduction:

The Daresbury experiment [27] gives $\phi_{\pi^+\pi^-} = 104° \pm 5°$.

The Desy-MIT one [28] gives $\phi_{\pi^+\pi^-} = 77° \pm 14°$ on the proton, $94° \pm 9°$ on Carbon and $105° \pm 15°$ on Pb.

A Rochester-Cornell experiment [29] gives $\phi_{\pi^+\pi^-} = 93° \pm 5°$ on C, $81° \pm 6°$ on Al and $81° \pm 7°$ on Pb.

A Slac experiment [30] gives $\phi_{\pi^+\pi^-} = 93° \pm 22$ on the proton. Roughly all measurements of $\phi_{\pi^+\pi^-}$ agree for a value near 90°. But in order to check $\varphi_{\varepsilon'}$ one has to substract $\phi_{e^+e^-}$ (see Eq. (16)). The Daresbury result gives $\varphi_{\varepsilon'}$ near zero whereas the Desy-MIT result is much larger and approach the theoretical value of 101°. Of course the large experimental errors should not be forgotten.

One may now discuss [32] these values of the production phases with respects to models of vector meson photoproduction. The V.D.M. allows to relate these phases to those of the vector meson elastic scattering:

$$\langle \tilde{V}C|B\gamma\rangle = \langle \tilde{V}C|B\varrho\rangle \langle \tilde{\varrho}|\gamma\rangle/m_\varrho^2 + \langle \tilde{V}C|B\omega\rangle \langle \tilde{\omega}|\gamma\rangle/m_\omega^2.$$

From relations (1) it follows in first order:

$$\langle \tilde{\varrho} C | B\varrho \rangle = A_{\varrho^0} - \varepsilon(A_{\varrho^0 \omega^0} + A_{\omega^0 \varrho^0}) \quad \langle \tilde{\omega} C | B\omega \rangle = A_{\omega^0} + \varepsilon(A_{\varrho^0 \omega^0} + A_{\omega^0 \varrho^0})$$

$$\langle \tilde{\varrho} C | B\omega \rangle = A_{\varrho^0 \omega^0} + \varepsilon(A_{\varrho^0} - A_{\omega^0}) \qquad \langle \tilde{\omega} C | B\varrho \rangle = A_{\omega^0 \varrho^0} + \varepsilon(A_{\varrho^0} - A_{\omega^0}).$$

In the case of elastic scattering $(B \equiv C)$ $A_{\varrho^0 \omega^0} = A_{\omega^0 \varrho^0} \equiv A_T$ and if isospin zero exchange is dominant $|A_T| \ll |A_{V^0}|$, then: $\langle \tilde{\varrho} C | B\varrho \rangle \simeq A_{\varrho^0} \langle \tilde{\omega} C | B\omega \rangle \simeq A_{\omega^0}$ first order included; but $\langle \tilde{\varrho} C | B\omega \rangle$ and $\langle \tilde{\omega} C | B\varrho \rangle$ are weaker (first order in A_T/A_{V_0} or in ε). One may then write:

$$\phi_P = \mathrm{Arg} \left\{ \frac{(A_{\omega^0} + 2\varepsilon A_T) f_{\omega\gamma} + (A_T + \varepsilon(A_{\varrho^0} - A_{\omega^0})) f_{\varrho\gamma}}{(A_{\varrho^0} - 2\varepsilon A_T) f_{\varrho\gamma} + (A_T + \varepsilon(A_{\varrho^0} - A_{\omega^0})) f_{\omega\gamma}} \right\}$$

$$\phi_P = \phi_0 + \mathrm{Im} \left\{ \frac{A_T + \varepsilon A_{\varrho^0}}{A_{\omega^0}} \cdot \frac{f_{\varrho^0 \omega}}{f_{\omega^0 \gamma}} - \frac{A_T - \varepsilon A_{\omega^0}}{A_{\varrho^0}} \cdot \frac{f_{\omega^0 \gamma}}{f_{\varrho^0 \gamma}} \right\} \equiv \phi_0 + \delta\phi_0 \tag{17}$$

with $\phi_0 \equiv \mathrm{Arg}(A_{\omega^0}/A_{\varrho^0})$ the relative phase of isospin states scattering. Let us discuss two situations:

a) If $|A_T| \ll |\varepsilon A_0|$, as ε is essentially purely imaginary,

$$\delta\phi_0 \lesssim |\varepsilon| \, |A_{\varrho^0} f_{\varrho^0 \gamma}/A_{\omega^0} f_{\omega^0 \gamma} + A_{\omega^0} f_{\omega^0 \gamma}/A_{\varrho^0} f_{\varrho^0 \gamma}|.$$

As the quark model predicts $|A_{\varrho^0}| \simeq |A_{\omega^0}|$ one expects (see Eq. (9)):

$$\delta\phi_0 \lesssim \varphi_{\omega\gamma} - \varphi_{\varrho\gamma} = 11°$$

when $A_{\varrho^0} = A_{\omega^0}$ also in phase, then $\varphi_0 = 0$ and $\phi_P = \delta\phi_0 = 11°$ exactly.

b) If $|A_T/A_{V_0}|$ is not smaller than $|\varepsilon|$ (i.e. if the $I \neq 0$ exchange is not smaller than 5% of $I = 0$ exchange), then $\delta\phi_0$ can be more important (see Eq. (17)).

It appears that the phase predictions of this model are well defined only if ϱ and ω are diffractively produced, then $\phi_P = \delta\phi_0 = 11°$ in this case.

The lepton pairs photoproduction measures also the individual phases ϕ_P^ω and ϕ_P^ϱ of the production amplitudes by their interferences with the real Bethe-Heitler term (Fig. 9).

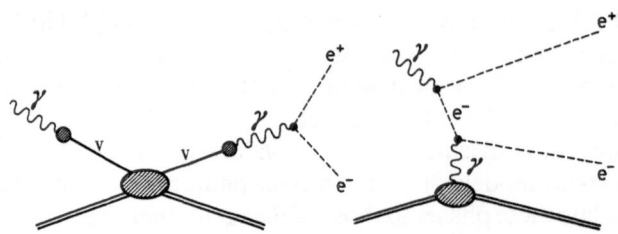

Fig. 9. Diagrams for lepton pairs photoproduction

The Desy-MIT results are:

$$\varphi_{\varrho\gamma} + \phi_P^\varrho = (11.8 \pm 4.4°) + \pi/2 \quad \text{and} \quad \varphi_{\omega\gamma} + \phi_P^\omega = (53° \pm 21°) + \pi/2.$$

The Daresbury ones are:

$$\varphi_{\varrho\gamma} + \phi_p^\varrho = (16.5 \pm 6.2°) + \pi/2 \quad \text{and} \quad \varphi_{\omega\gamma} + \phi_p^\omega = (116° \pm 22°) + \pi/2$$

The ϱ phase is compatible with a diffractive process $(\pi/2 + 2\varphi_{\varrho\gamma} \simeq \pi/2 - 2°)$, but for the ω one it would be expected for a diffractive process:

$$\varphi_{\omega\gamma} + \phi_p^\omega = \pi/2 + 2\varphi_{\omega\gamma} \simeq \pi/2 + 20°$$

which could be compatible with the Desy result but not with the Daresbury one.

A large non diffractive ω production (for example with π-exchange) could explain the large ϕ_p^ω value but this would not solve the problem for $\varphi_{\varepsilon'}$ when the production phases are eliminated by Eq. (16) in the case of the Daresbury results.

Clearly more experimental results are needed, in particular on the proton, in order to remove possible unknown nuclear effects.

VII. Other Processes Involving Vector Mesons

Any strong production of vector mesons $A + B \rightarrow C + V_{\rightarrow F}$ can be analyzed on the same footing; from the general amplitude:

$$R_{fi} = \frac{\langle F|\varrho\rangle \langle \tilde{\varrho}C|AB\rangle}{s - m_\varrho^2 + im_\varrho\Gamma_\varrho} + \frac{\langle F|\omega\rangle \langle \tilde{\omega}C|AB\rangle}{s - m_\omega^2 + im_\omega\Gamma_\omega} + R_B \quad (18)$$

an interference term will appear in the region $s \simeq m_\omega^2$. However the nature of this interference depends upon the production phases in $\langle \tilde{\omega}C|AB\rangle$ and $\langle \tilde{\varrho}C|AB\rangle$. They vary with the nature of the particles A, B and C and with the kinematical conditions. From the cross-section (see Appendix B):

$$\begin{aligned}
d\sigma/ds = C_F(s) \{&|\sqrt{\sigma^\varrho(s)}\, g_\varrho F/\Delta_\varrho(s) \\
&+ ce^{i\phi_P} \sqrt{\sigma^\omega(s)}\, g_{\omega F}/\Delta_\omega(s)|^2 + (1 - c^2)\, \sigma^\omega(s)\, |g_{\omega F}/\Delta_\omega(s)|^2\}
\end{aligned} \quad (19)$$

no general shape (see Fig. 10) of the $s = p_V^2 = p_F^2$ distribution can be predicted without a definite model for the production amplitudes (phase ϕ_p and degree of coherence c) or without an independant measurement of them.

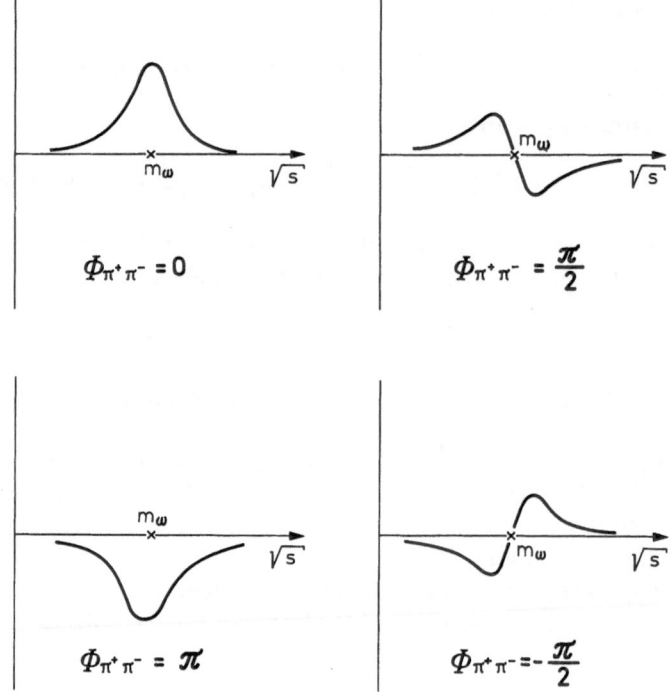

Fig. 10. Interference shapes as a function of the total phase ϕ

In fact an ideal phenomenological program would be once more to determine first these quantities c and ϕ_p from the lepton pairs production $A + B \to C + V_{\to e^+ e^-}$.

In this case $\phi_{e^+ e^-} \equiv \mathrm{Arg} \left\{ \dfrac{\langle \gamma | \omega \rangle \langle \tilde{\omega} C | AB \rangle}{\langle \gamma | \varrho \rangle \langle \tilde{\varrho} C | AB \rangle} \right\} = \phi_p + \varphi_{\omega\gamma} - \varphi_{\varrho\gamma}$.

Then the interference shape of an hadronic, for example $A + B \to C + V_{\to \pi^+ \pi^-}$ will be determined:

$$\phi_{\pi^+ \pi^-} \equiv \mathrm{Arg} \left\{ \frac{\langle 2\pi | \omega \rangle \langle \tilde{\omega} C | AB \rangle}{\langle 2\pi | \varrho \rangle \langle \tilde{\varrho} C | AB \rangle} \right\} = \phi_p + \varphi_{\varepsilon'},$$

and $\phi_{\pi^+ \pi^-} - \phi_{e^+ e^-} = \varphi_{\varepsilon'} - \varphi_{\omega\gamma} + \varphi_{\varrho\gamma}$ could be checked.

Up to now this method has not been experimentally used for strong production of vector mesons and one has to rely on models if one wants to understand the various $\varrho - \omega$ interference effects seen in $\pi^+ \pi^-$ production. This subject has already been reviewed by Roos [33] and Marshall [31].

Table 1

Reaction	Reference	$R(\%)$	$\phi(°)$	ϕ_p (theory)	$\phi_{\varepsilon'}$ (deduced)
$e^+e^- \to \pi^+\pi^-$	[22]	$3.3^{+2.1}_{-1.6}$	$89° \pm 15°$	$11°$	$78° \pm 15°$
$\gamma + C \to C + \pi^+\pi^-$	[27]	>0.8	$104° \pm 5°$		
$\gamma + C \to C + \pi^+\pi^-$	[28]	>0.95	$94° \pm 9°$		
$\gamma + C \to C + \pi^+\pi^-$	[29]	>1.2	$93° \pm 5°$		
$\gamma + Pb \to Pb + \pi^+\pi^-$	[28]	>1.2	$105° \pm 15°$		
$\gamma + Pb \to Pb + \pi^+\pi^-$	[29]	>1.6	$81° \pm 7°$		
$\gamma + Al \to Al + \pi^+\pi^-$	[29]	>1.4	$81° \pm 6°$		
$\gamma + p \to p + \pi^+\pi^-$	[28]	>1.8	$77° \pm 14°$		
$\gamma + p \to p + \pi^+\pi^-$	[30]		$93° \pm 22°$		
$\pi^+ p \to \Delta^{++} + \pi^+\pi^-$	[37]	>0.25	$192° \pm 17°$	$90°$	$102° \pm 17°$
$\pi^+ p \to \Delta^{++} + \pi^+\pi^-$	[38]		$163° \pm 23°$	$90°$	$73° \pm 23°$
$\pi^+ p \to \Delta^{++} + \pi^+\pi^-$	[39]		$\simeq 90°$		
$\pi^- p \to \pi^+\pi^- n$	[40]	>0.36	-15 ± 30	$-90°$	$75° \pm 30°$
$\pi^- p \to \pi^+\pi^- n$	[41]		$-4° \pm 20°$	$-90°$	$86° \pm 30°$
$\pi^+ n \to \pi^+\pi^- p$	[42]		(around $180°$)	$90°$	
$K\bar{p} \to \Lambda + \pi^+ + \pi^-$	[43]	>0.2			
$\bar{p}p \to 2\pi^+ 2\pi^-$	[44]	>1			
$\bar{p}p \to 2\pi^+ 2\pi^-$	[45]	>1.9			
$\pi^- p \to p + \pi^+\pi^-\pi^-$	[46]	>1.1			
$\bar{p}n \to 3\pi, 4\pi, 5\pi, 6\pi$	[47]	<4.3			

For π-Nucleon reactions a model based on the exchange degeneracy of the π and B Regge trajectories has been proposed [8]. This gives the relative phase ϕ_p at the production, and knowing ε, gives the interference shape. A qualitative agreement arises for $\pi^+ p \to \pi^+\pi^- \Delta^{++}$, $\pi^- p \to \pi^+\pi^- n$ and perhaps in $\pi^+ n \to \pi^+\pi^- p$ and $K^- p \to \pi^+\pi^- \Lambda$ (see Table 1).

The $\bar{p}p \to$ pions reactions have also revealed $\varrho - \omega$ interference effects but it is difficult in this case to propose a value for the production phases.

Concerning the intensity of the effects in all these strong productions, it is also difficult to deduce from it the value of the $\omega \to 2\pi$ branching ratio due to the presence of the coherence factor c; as $0 \leq c \leq 1$ only a lower limit of the branching ratio can be given in general in a model independant way (see Table 1).

Finally the $\bar{p}n$ annihilation at rest is a privileged reaction; as $I = 1$ is fixed and as the annihilation occurs only in $L = 0$ states, G parity is related to the total angular momentum $G \equiv (-1)^{J+1}$ and the ϱ^0 and ω^0 productions are incoherent. In such a case the mixing effect is only proportional to $|\varepsilon|^2$ giving a small ω peak in the $\pi^+\pi^-$ production. This has been used to set an upper limit on the $\omega \to 2\pi$ branching ratio (see Table 1).

VIII. Conclusion

The essential features of the $\varrho - \omega$ mixing problem seem to be presently understood. The dominant effects come from a mixing term ε which is related to a $|\varDelta I| = 1$ E.M. transition similar to the E.M. mass differences of the mesons. The near degeneracy of the ϱ and ω masses reinforces the effects, $|\varepsilon| \simeq 8\,\alpha$. As a consequence, and essentially in $\omega \to 2\pi$, but also in $\varrho \to 3\pi$ the mixing term dominates over a possible direct E.M. transition $\omega^0 \to 2\pi$ or $\varrho^0 \to 3\pi$. This leads to precise phase predictions which are now in agreement with the last experimental results for $e^+ e^- \to 2\pi$. Similar predictions in magnitude and in phase are made for the other decay modes of the ϱ and ω mesons. The reaction $e^+ e^- \to 3\pi$ is only slightly modified due to a broad ϱ background under the narrow ω peak. The $\omega - \gamma$ and the $\varrho - \pi - \gamma$ couplings get a phase of about $10°$.

Concerning the reactions with production of vector mesons (photoproduction or strong production) the point under doubt is the phase of the production amplitudes and their degree of coherence. The $\varrho - \omega$ interference shape depends upon this as well as upon the $\varrho - \omega$ mixing phase. Much experimental works for example on lepton pair production and on 2π production have to be done before one reachs a definite conclusion. Nevertheless in various reactions the magnitude of the effects already observed is in agreement with the predicted value for $\varGamma_{\omega \to 2\pi}$.

If one imagines that the experimental results could reach a more accurate level, then other interesting features could be considered.

First if the 3π decay modes could be analyzed in order to reach the $f_{\varrho, 3\pi}$ coupling, then a test of the mixing formalism and of the dominance of the mixing terms over the direct transitions could be done with the help of Eq. (8).

If the weakness of the $\varrho - \varphi$ mixing could be confirmed or eventually its magnitude measured, then one would have one more information about the E.M. self-masses in the vector meson nonet. This would complete the problem of the $|\varDelta I| = 1$ mass differences which is simultaneously a problem of invariance and of connections between low energy and high energy behaviour of Compton scattering amplitudes.

A very precise phase determination of the couplings of the ϱ and ω mesons could say something on the type of mixing (current-mixing or mass-mixing). The $\varrho - \omega$ case is however not very favourable for such a discrimination; an accuracy of a few degrees is required and at this level the finite width effects [34] of the ϱ also occur (i.e. the usual Breit-Wigner formula should be modified). From a purely theoritical point of view current mixing is more likely as it preserves the charges of the particles [7].

In any case the merit of the $\varrho - \omega$ mixing effects will be also once more the possibility of using the well known E.M.I. for looking inside the hadronic properties. The various $\varrho - \omega$ interference effects in production reactions will be used for determining the relative phases of the strong production amplitudes; this will test many high energy models[1].

$\varrho - \omega$ mixing appears as one of the scarce small effects at high energy where detailed fits of experimental results let us learn something in physics.

Appendix A.
General Formulas for Mass- and Current Mixing

These are essentially the results of Sachs and Willemsen [10]. One writes $F(s) = [s\pi(s) - M_0^2]^{-1} = \pi(s)^{-\frac{1}{2}}[s - W(s)]^{-1}\pi(s)^{-\frac{1}{2}}$, where $W(s) \equiv \pi(s)^{-\frac{1}{2}}M_0^2\pi(s)^{-\frac{1}{2}}$ and $M_0^{-2} \equiv \int ds' \varrho(s')/s'$ (real). From unitarity and analyticity: $\operatorname{Im} F^{-1}(s) = s \operatorname{Im} \pi(s) = s^{-\frac{1}{2}}\Gamma(s)$, then in the Breit-Wigner approximation:

$$\pi(s) = \begin{pmatrix} 1 + i\Gamma_\varrho/m_\varrho & 2\lambda \\ 2\lambda & 1 + i\Gamma_\omega/m_\omega \end{pmatrix} \quad \text{and} \quad M_0^2 = \begin{pmatrix} m_\varrho^2 & \eta m_\varrho^2 \\ \eta m_\varrho^2 & m_\omega^2 \end{pmatrix}.$$

η is real, but λ is complex (its imaginary part satisfies a unitarity relation as does δm_I defined in Section III).

By diagonalizing W and renormalizing with $\pi^{-\frac{1}{2}}(s)$, one gets the physical states:

$$\begin{aligned} |\varrho\rangle &= e^{-i\theta}|\varrho^0\rangle - \varepsilon_1 |\omega^0\rangle \\ |\omega\rangle &= \varepsilon_2 |\varrho^0\rangle + |\omega^0\rangle \end{aligned} \tag{A.1}$$

with $\theta \equiv \Gamma_\varrho/2m_\varrho$ (Γ_ω/m_ω has been neglected) and $\varepsilon_1 = s + \lambda e^{-i\theta}$, $\varepsilon_2 = (s - \lambda) e^{-i\theta}$, in which $s \equiv \dfrac{e^{i\theta} m_\varrho^2 [\eta - 2\lambda e^{-2i\theta}]}{m_\omega^2 - m_\varrho^2 - i(m_\omega \Gamma_\omega - m_\varrho \Gamma_\varrho)}$.

This is the general case expressed in terms of three parameters η, $\operatorname{Re}\lambda$ and $\operatorname{Im}\lambda$. If $\lambda \equiv 0$ one has the real mass-mixing case and if $\eta \equiv 0$ one has pure current-mixing. One verifies however that up to small phase

---

[1] It has been asserted [35] that $\varrho - \omega$ interference could be used to determine the sign of the $\omega - \varphi$ mixing angle θ_V, for which broken SU_3 gives only the modulus $|\theta_V| \simeq 40°$. These people forget that $\sin \theta_V$ appears twice in a $\pi^+ \pi^-$ production amplitude; a first time in the ω production amplitude (for ex. $3^{-\frac{1}{2}} f_{\varrho\gamma} \sin \theta_V$ for $e^+ e^-$ or $3^{-\frac{1}{2}} A_\varrho \sin \theta_V$ for strong production) and a second time in the $\omega \to 2\pi$ coupling (i.e. in ε' through $\delta m_R \propto (\sin \theta_V)^{-1}$; see Section III and Eq. (6)).

This is for example obvious in the VDM contribution [14] which depends upon $f_{\omega\gamma}^2$ (i.e. $\sin^2 \theta_V$); then, one knows that all the other signs are correlated to the one of this contribution independantly of any convention [36]. It is therefore not possible to determine the sign of $\sin \theta_V$ from these effects.

deviations ($\theta \simeq 4°$) the phenomenological situation is similar to the one obtained in Section II and III:

s is equivalent to the quantity ε defined by Eq. (2) in Section II and expressed in terms of the complex quantity $W_{\varrho^0\omega^0} \equiv (m_\varrho + m_\omega)(\delta m_R + i\delta m_I)$.

λ/s is of the order of $\Gamma_\varrho/2m_\varrho$ and should be consistently neglected; then $\varepsilon_1 \simeq \varepsilon_2 \simeq s \simeq \varepsilon$, and the mixing relations (A.1) are identical to the relations (1) of Section II.

In order to distinguish between different kinds of mixing (η or $\mathrm{Re}\,\lambda$) it is therefore necessary to measure phases of $4°$. This is presently out of the experimental accuracy.

Appendix B.
Helicity Dependence of Vector Meson Productions

The amplitude corresponding to the process $A + B \to C + V_{\to F}$ may be written:

$$R_{fi} = \sum_{\alpha,\lambda} \frac{\langle F|\alpha(\lambda)\rangle \langle \tilde{\alpha}(\lambda)\,C|AB\rangle}{s - m_\alpha^2 + im_\alpha\Gamma_\alpha}$$

where $\alpha \equiv \omega, \varrho$ and λ is the helicity state of the vector meson. One can in the usual cases ($F \equiv 2\pi, 3\pi, \pi\gamma, e^+e^- \ldots$) factorize a coupling constant: $\langle F|\alpha(\lambda)\rangle \equiv f_{\alpha F} a_F(\lambda)$, and define through the amplitude $a_F(\lambda)$ the density matrix:

$$\varrho_F(\lambda, \lambda') = a_F(\lambda)\, a_F^*(\lambda')/\sum_\lambda |a_F(\lambda)|^2 .$$

Similarly one defines also production density matrices:

$$D_{(\lambda,\lambda')}^{\pm\,\alpha,\beta} \equiv \frac{\displaystyle\sum_{ABC} [\langle \tilde{\alpha}(\lambda)\,C|AB\rangle \langle \tilde{\beta}(\lambda')\,C|AB\rangle^* \pm \langle \beta(\lambda)\,C|AB\rangle \langle \tilde{\alpha}(\lambda')\,C|AB\rangle^*]}{\displaystyle\sum_{ABC}\sum_\lambda [\langle \tilde{\alpha}(\lambda)\,C|AB\rangle \langle \tilde{\beta}(\lambda)\,C|AB\rangle^* \pm \langle \tilde{\beta}(\lambda)\,C|AB\rangle \langle \tilde{\alpha}(\lambda)\,C|AB\rangle^*]}$$

$\left(\displaystyle\sum_{ABC} \text{means summations over the spins of the particles } A, B \text{ and } C\right)$. Then, through $\Sigma|R_{fi}|^2$ one will get the cross-section:

$$d\sigma/ds \propto \sum_{\alpha,\beta} \mathrm{Tr}(\varrho_F D^{+\alpha\beta})\,\mathrm{Re}(\sigma^{\alpha\beta}(s))\,[f_{\alpha F} f_{\beta F}^*/\Delta_\alpha(s)\,\Delta_\beta^*(s)]$$

$$+ \sum_{\alpha,\beta} \mathrm{Tr}(\varrho_F D^{-\alpha\beta})\,i\,\mathrm{Im}(\sigma^{\alpha\beta}(s))\,[f_{\alpha F} f_{\beta F}^*/\Delta_\alpha(s)\,\Delta_\beta^*(s)]$$

in which $\Delta_\alpha(s) \equiv s - m_\alpha^2 + im_\alpha\Gamma_\alpha$

$$\mathrm{Re}\,\sigma^{\alpha\beta}(s) \propto \sum_{ABC}\sum_\lambda [\langle \tilde{\alpha}(\lambda)\,C|AB\rangle \langle \tilde{\beta}(\lambda)\,C|AB\rangle^*$$

$$+ \langle \tilde{\beta}(\lambda)\,C|AB\rangle \langle \tilde{\alpha}(\lambda)\,C|AB\rangle^*],$$

$$\mathrm{Im}\,\sigma^{\alpha\beta}(s) \propto (-i) \sum_{ABC}\sum_\lambda [\langle \tilde{\alpha}(\lambda)\,C|AB\rangle \langle \tilde{\beta}(\lambda)\,C|AB\rangle^*$$

$$- \langle \tilde{\beta}(\lambda)\,C|AB\rangle \langle \tilde{\alpha}(\lambda)\,C|AB\rangle^*],$$

such that $\sigma^{\alpha\beta}(\lambda, \lambda') = D^{+\alpha\beta}(\lambda, \lambda') \operatorname{Re}\sigma^{\alpha\beta}(s) + i\bar{D}^{\alpha\beta}(\lambda, \lambda') \operatorname{Im}\sigma^{\alpha\beta}(s)$ is a "mixed cross-section" related to $\langle \tilde{\alpha}(\lambda) C | AB \rangle \langle \tilde{\beta}(\lambda') C | AB \rangle^*$; ($\sigma^{\alpha\alpha}(\lambda, \lambda)$ is exactly the production cross-section of a vector meson $\alpha(\lambda)$). Then, after integration over angles:

$$d\sigma/ds = C_F(s)\left\{ \sum_{\alpha = \varrho, \omega} (\sigma^\alpha(s)|f_{\alpha F}/\Delta_\alpha(s)|^2) \right.$$
$$+ 2\operatorname{Re}\sigma^{\varrho\omega}(s)\operatorname{Re}(f_{\varrho F}f_{\omega F}^*/\Delta_\varrho(s)\,\Delta_\omega^*(s))$$
$$\left. - 2\operatorname{Im}\sigma^{\varrho\omega}(s)\operatorname{Im}[f_{\varrho F}f_{\alpha F}^*/\Delta_\varrho(s)\,\Delta_\omega^*(s)]\right\}$$

in which $C_F(s)$ is a kinematical factor.

The last two terms are often written in terms of a factor of coherence c; if ϕ_λ is the relative phase of $\langle \tilde{\omega}(\lambda) C | AB \rangle$ with respects to $\langle \tilde{\varrho}(\lambda) C | AB \rangle$, then:

$$\operatorname{Re}\sigma^{\varrho\omega}(s) \propto \sum_\lambda 2\cos\phi_\lambda |\langle \tilde{\omega}(\lambda) C | AB \rangle \langle \tilde{\varrho}(\lambda) C | AB \rangle| \equiv 2c\cos\phi_p \sqrt{\sigma^\varrho(s)\sigma^\omega(s)}$$
$$\operatorname{Im}\sigma^{\varrho\omega}(s) \propto \sum_\lambda -2\sin\phi_\lambda |\langle \tilde{\omega}(\lambda) C | AB \rangle \langle \tilde{\varrho}(\lambda) C | AB \rangle|$$
$$= -2c\sin\phi_p \sqrt{\sigma^\varrho(s)\,\sigma^\omega(s)}$$

where $0 \leq c \leq 1$, and ϕ_p is an "overall production phase".

The complete cross-section will then be written.

$$d\sigma/ds = C_F(s)\,\{|\sqrt{\sigma^\varrho(s)}\,f_{\varrho F}/\Delta_\varrho(s) + ce^{i\phi_p}\sqrt{\sigma^\omega(s)}\,f_{\omega F}/\Delta_\omega(s)|^2\}$$
$$+ (1 - c^2)\,\sigma^\omega(s)\,|f_{\omega F}/\Delta_\omega(s)|^2 \tag{B.1}$$

In the case of $\pi^+\pi^-$ production:

$$d\sigma/ds \simeq C_{2\pi}(s)\,f_{\varrho\pi\pi}^2\{|\sqrt{\sigma^\varrho(s)}/\Delta_\varrho(s) + c\varepsilon'\,e^{i\phi_p}\sqrt{\sigma^\omega(s)}/\Delta_\omega(s)|^2$$
$$+ (1 - c^2)\,|\varepsilon'|^2\,\sigma^\omega(s)/|\Delta_\omega(s)|^2\}\,.$$

In general, as c is not independantly measured, a fit of $d\sigma/ds$ gives only the phase $\phi_p + \varphi_{\varepsilon'}$ and the product $c|\varepsilon'|$; as $0 \leq c \leq 1$ only a lower limit can be given for $|\varepsilon'|$ and then for $\Gamma_{\omega \to 2\pi}$.

The above formulas are expressed in terms of the physical production amplitudes, but it is easy with the linear combinations (1) of Section II to make appear the ones for isospin states ϱ^0 and ω^0.

References

1. Glashow, S.: Phys. Rev. Letters **7**, 469 (1961).
2. Wigner, E. P., Weisskopf, V. F.: Z. Physik **63**, 54 (1930); **67**, 18 (1930).
3. Bernstein, J., Feinberg, G.: N.C. **27**, 1343 (1962).
4. Horn, D.: Phys. Rev. D1, 1421 (1970).
5. Stodolsky, L.: Phys. Rev. 5D (1970).
6. Jacob, R., Sachs, R. G.: Phys. Rev. **121**, 350 (1961); – Harte, J., Sachs, R. G.: Phys. Rev. **135**, B 459 (1964).
7. Coleman, S., Schnitzer, H. J.: Phys. Rev. **134**, B 863 (1964).

8. Goldhaber, A. S., Fox, G. C., Quigg, C.: Phys. Letters **30** B, 249 (1969).
9. Gourdin, M., Renard, F. M., Stodolsky, L.: Phys. Letters **30** B, 347 (1969).
10. Sachs, R. G., Willemsen, J. F.: Phys. Rev. D **2**, 133 (1970); – Willemsen, J. F.: preprint EFI 70–11 (1970).
11. Gourdin, M.: Boulder Lectures (1969).
12. —Invited talk at the Bologna Conference, April 1971.
13. Renard, F. M.: Nucl. Phys. B **15**, 118 (1970).
14. Nambu, Y., Sakurai, J.: Phys. Rev. Letters **8**, 79 (1962); – Gell-mann, M., Sharp, D., Wagner, W. G.: Phys. Rev. Letters **8**, 261 (1962); – Gatto, R.: N.C. **28**, 658 (1963); – Singer, P.: Phys. Rev. Letters **12**, 524 (1964).
15. Coleman, S., Glashow, S. L.: Phys. Rev. **134**, B 671 (1964).
16. Particle Data, April 1971.
17. Gourdin, M.: Unitary Symmetries, North Holland (1967).
18. Dietz, K., Krauze, F., Rollnik, H., Pietschmann, P.: Contribution to the Vienna Conference (1968).
19. Gal, A., Scheck, F.: Nucl. Phys. B **2**, 110 (1967).
20. Mathur, V. S., Okubo, S.: preprint Rochester UR–875–280.
21. Perez-Y-Jorba, J.: Daresbury study weekend (1970).
22. Parrour, G., et al.: Contribution to the Bologna Conference April 1971.
23. Gourdin, M., Renard, F. M.: unpublished.
24. Biggs, P. J., et al.: Phys. Rev. Letters **24**, 1197 (1970).
25. — Daresbury preprint DNPL/P 70 (1970).
26. Alvensleben, H., et al.: Phys. Rev. Letters **25**, 1373 and 1377 (1970); Nucl. Phys. B **25**, 333 and 342 (1971).
27. Biggs, P. J., et al.: Phys. Rev. Letters **26**, 1201 (1970).
28. Alvensleben, H., et al.: Desy preprint (1971).
29. Behrend, H. J., et al.: Contribution to the Bologna Conference April 1971.
30. Moffeit, K. C., et al.: UCRL 19753 (1970).
31. Marshall, R.: Invited talk at the Bologna Conference April 1971.
32. Gourdin, M.: Daresbury study weekend (1970).
33. Roos, M.: Daresbury study weekend (1970).
34. Gounaris, G. J., Sakurai, J. J.: Phys. Rev. Letters **21**, 244 (1968); – Gounaris, G. J.: Phys. Rev. **181**, 2066 (1969); – Renard, F. M.: Nucl. Phys. B **15**, 267 (1970).
35. Parsons, R. G.: Phys. Rev. **188**, 2277 (1969); – Goldhaber, A. S., Fox, G. C., Quigg, C.: Phys. Letters **30** B, 249 (1969); – Dass, G. V.: Daresbury study weekend (1970).
36. Allcock, G. R.: Nucl. Phys. B **21**, 269 (1970).
37. Goldhaber, G., et al.: Phys. Rev. Letters **23**, 1351 (1969).
38. Jackson, W. C., et al.: Bull. Ann. Phys. Soc. **15**, 513 (1970).
39. Flatte, S., et al.: Contribution to the Bologna Conference April (1971).
40. Hagopian, S., et al.: Florida State preprint FSU-HEP-70-8-1 (1970).
41. Dalpiaz, P., et al.: Contribution to the Bologna Conference April (1971).
42. Katz, P., et al.: Bull. Ann. Phys. Soc. **15**, 69 EG 7 (1970) a reference given by D. G. Coyne, UCRL 19450 (1970).
43. Flatte, S.: Phys. Rev. **1** D, 1 (1970).
44. Allison, W., et al.: Phys. Rev. Letters **24**, 618 (1970).
45. Chapman, J., et al.: Michigan preprint (1970).
46. Diaz, J., et al.: Nucl. Phys. B **16**, 239 (1970).
47. Bizarri, R., et al.: preprint Roma (1970).

Dr. F. M. Renard
Department de Physique Mathématique
Faculté des Sciences
34 Montpellier/France

Exotic Electromagnetic Currents

A. Donnachie

Contents

1. Introduction

That the strong interactions exhibit an approximate $SU(3) \times SU(3)$ symmetry implies the existence of a corresponding set of conserved or approximately conserved currents to which the octets of vector and axial vector currents which mediate the electromagnetic and weak interactions seem to be closely related. It is natural to query the ways in which this symmetry may be broken and one way, which is the one we shall consider here, is that there may exist additional weak or electromagnetic currents lying outside $SU(3)$ – exotic currents. Such currents are not directly relevant either to strong interaction considerations or to $SU(2) \times SU(2)$ results. However in some current algebra applications to weak and electromagnetic processes their absence is explicitly assumed and it is in some of these processes, particularly K_{13} and $\eta \rightarrow 3\pi$ decays that the most severe difficulties have arisen.

We shall concern ourselves exclusively with exotic electromagnetic currents, and consider two particular aspects of exotic behaviour, namely that the electromagnetic current has a C-violating component and that either or both of the C-conserving and C-violating currents have an isotensor, as well as the usual isoscalar and isovector components. A C-violating electromagnetic current was proposed by Bernstein, Feinberg and Lee [1] as an explanation of the CP violation observed in K^0 decay. The magnitude of the branching ratio $\Gamma(K_L \rightarrow \pi^+ \pi^-)/\Gamma(K_L \rightarrow 3\pi)$ $\simeq \alpha/\pi$ leads to the conjecture that the CP violation is caused by the electromagnetic correction to the CP conserving weak decay. The possibility of an isotensor electromagnetic current was first raised by Grishin, Lyuboshitz, Ogievetskii and Podgoretskii [2] and by Dombey and Kabir [3]. The essence of their argument was that while data showed that isoscalar and isovector components were necessary parts of an electromagnetic current, none of it specifically excluded an isotensor (or higher) component. Not only were existing data not sensitive to the presence of an isotensor component, very few unambiguous tests could be proposed. One which was suggested was to study the radiative decays of the charged and neutral states of an $I = \frac{3}{2}$ resonance, the $P_{33}(1236)$ in particular. This latter point was elaborated initially by Shaw [4] and later by Sanda and Shaw [5].

We shall decompose the C-conserving electromagnetic current J_μ and the C-violating electromagnetic current K_μ into their various isospin components, so that

$$
\begin{aligned}
J_\mu &= J_\mu^0 + J_\mu^1 + J_\mu^2 \\
K_\mu &= K_\mu^0 + K_\mu^1 + K_\mu^2
\end{aligned}
\tag{1.1}
$$

where the superscripts 0, 1, 2 denote isoscalar, isovector, isotensor respectively. Obviously the charge conjugation properties of these currents are

$$
\begin{aligned}
C^{-1} J_\mu^i C &= -J_\mu^i \\
C^{-1} K_\mu^i C &= K_\mu^i
\end{aligned}
\qquad i = 0, 1, 2.
\tag{1.2}
$$

Now the principle of minimal electromagnetic interactions allows one to deduce the form of the electromagnetic interaction, and we know it does it correctly in the case of spin-$\frac{1}{2}$ leptons. What does it say in general?

For spin-0 and spin-$\frac{1}{2}$ particles it is well known to specify uniquely the form of the interaction in terms of the charge, the current transforming as the charge. Defining the charges

$$
Q^i = \int J_0^i(x)\, d^3x \qquad \tilde{Q}^i = \int K_0^i(x)\, d^3x \,.
\tag{1.3}
$$

Lee [6] has shown that since Q^i and $\tilde{Q}^i$ are conserved quantum numbers whose eigenvalues are additive, $Q^i (i \neq 1)$ and $\tilde{Q}_i$ are isoscalar. Hence the currents for spin-0 and spin-$\frac{1}{2}$ fields must necessarily be isoscalar, i.e. $K_\mu^i = 0$ for $i \neq 0$ and $J_\mu^i = 0$ for $i \neq 0$ or 1 and there is no room for isotensor currents.

However the case of higher spin is quite different. Lee [7] has studied spin-1 fields in detail, and has shown that in this case the current need not transform as the charge and there is no reason to restrict its isospin transformation properties. As an example, paraphrasing an argument due initially to Adler [8], consider vector and axial vector fields ψ_μ and

ψ_μ^5 respectively, transforming in the same way under C and let ϕ_μ^5 be an axial vector field with the opposite transformation properties under C. Then the currents

$$J_\mu = \varepsilon_{\mu\nu\gamma\delta}\, \partial_\nu \psi_\gamma \psi_\delta^5$$
$$K_\mu = \varepsilon_{\mu\nu\gamma\delta}\, \partial_\nu \psi_\gamma \phi_\delta^5 \tag{1.4}$$

have zero charge and choosing suitable isospin properties for ψ_μ, ψ_μ^5 and ϕ_μ^5 we can construct J_μ and K_μ with any isospin properties we choose.

To summarize, the principle of minimal electromagnetic interactions uniquely specifies the form of the interaction for spin-0 and spin-$\frac{1}{2}$ particles, the current depending only on the charge. Any C-violating current thus constructed must necessarily be isoscalar and the C-conserving current can only be isoscalar and isovector. For higher spin particles, the result is no longer unique, parameters other than the charge are needed to fix the form of the interaction, and exotic isospin and C-violation with any isospin are allowed. These terms enter in three point vertices ($\gamma A B$) which can not be derived from the charge and must be viewed, at least at present, as fundamental couplings which can only be obtained from experiment.

The features to look for are the presence of high isospin and high spin, both of which can be found conveniently in the $\varDelta N \gamma$ vertex, already suggested in connection with the isotensor current alone. We shall consider in detail four situations in which this vertex plays a role, namely pion photoproduction and π-radiative capture in the first resonance region; deuteron photodisintegration and neutron-proton radiative capture; inelastic electron scattering off a polarized target in the region of the first resonance; the neutron electric dipole moment.

Before entering this discussion, we comment briefly on the evidence for exotic electromagnetic currents which can be deduced from other reactions. Boson decays have been studied extensively with this object in view. Since bosons lie in $SU(3)$ singlets and octets and have $I \leqq 1$, the possible couplings of an isotensor current are severely restricted, and since neutral bosons are eigenstates of C, even if $I = 2$ couplings are allowed only one of J_μ^2, K_μ^2 is possible. Further if we consider the 0^- nonet, the only $0^- 0^- \gamma$ coupling allowed by isospin for $I = 2(\pi\pi\gamma)$ is forbidden by Bose statistics for both $C = \pm 1$ and for $1^- 0^- \gamma$ the only one allowed by isospin ($\varrho\pi\gamma$) is forced by G-parity to be C-conserving. Coupling of K_μ^2 only becomes possible for vertices such as $A_2 \pi \gamma$, thus for first-order low-mass boson processes one would not expect to find any effects due to a possible K_μ^2 term. It is much more speculative to extend these conclusions to second order decays, since in general reasonable dynamical models are not available. There is, however, an obvious exception in $\mu \to 3\pi$ decay where the conventional terms are known to be heavily suppressed and K_μ^2 effects here might be relatively enhanced.

$\eta \to \pi^0 e^+ e^-$. The upper bound for the branching ratio for this decay is 0.01%. To first order in α, the decay can proceed only by K_μ' whose effect is therefore negligible.

$\eta \to \pi^+ \pi^- \pi^0$. From an analysis of 36 000 events, Gormley *et al.* [9] obtained an overall charge asymmetry of 1.5 ± 0.5% and a sextant asymmetry, which is sensitive to the charge asymmetry in the $I = 0$ final 3π state, of 0.5 ± 0.5%, indicating that any C-violation is most probably occurring in the $I = 2$ final 3π state.

The matrix element for the decay is proportional to

$$\int d^4 x \int d^4 q \, D_{\mu\nu}(q) \exp i q x \langle \mu | T(J_\mu^{\text{e.m.}}(x) \, J_\nu^{\text{e.m.}}(0)) | \pi^+ \pi^- \pi^0 \rangle$$

where $D_{\mu\nu}$ is the photon propagator and $J_\mu^{\text{e.m.}} \equiv J_\mu + K_\mu$.

Of the six possible combinations of currents which could at first sight lead to an $I = 2$ final 3π state, $K^1 J^2$ and $K^2 J^1$ can not contribute since they do not have, in fact, a component which transforms as $I = 2$. Additionally they have opposite G-parity. If we accept the conclusion drawn from $\eta \to \pi^0 e^+ e^-$ decay then we can rule out $K^1 J^1$ and if $K^0 J^2$ were responsible then we would expect at least as big an asymmetry from $K^0 J^0$ whose effect should be seen, barring a fortuitous cancellation, in the sextant asymmetry. If we then discard $K^2 J^2$ on the naive assumption that it should be small since it is doubly exotic, we are left with the conclusion that the asymmetry in this decay is caused by $K^2 J^0$.

We finally note that if an isotensor contribution is present, be it C-violating or otherwise, then the branching ratio

$$\Gamma(\eta \to 3\pi^0) / \Gamma(\eta \to \pi^+ \pi^- \pi^0)$$

should depart from the usual theoretical prediction of about 1.6. In a recent experiment, Bultram *et al.* [10] have obtained a value of 1.38 ± 0.09 for this ratio which is certainly consistent with the presence of an $I = 3$ final state.

$\eta \to \pi^+ \pi^- \gamma$. Observation of a charge asymmetry in this decay implies the presence of either K_μ^0 or K_μ^2 since the C-violating amplitude must have an $I = 0$ or $I = 2$ final pion state. Gormley *et al.* [11] have obtained a value of 2.4 ± 1.4% which is clearly inconclusive. However, due to angular momentum conservation, the lowest orbital angular momentum state for the pions is d-wave and these events are kinematically suppressed.

$K_L^0 \to \pi^+ \pi^-$. This is the only decay in which CP violation is firmly established and, as we have already discussed, lead to the conjecture that it is due to a C-violating electromagnetic correction to the CP conserving weak decay. Unfortunately in this model the matrix element for the decay involves the time-ordered product of four current operators (two weak, two electromagnetic) and the ensuing complexity has prevented any deduction of isospin selection rules for the overall process. Until this theoretical ambiguity can be clarified, no conclusion can be drawn.

$K^\pm \to \pi^\pm \pi^0 \pi^0$; $\pi^\pm \pi^0 \gamma$. Preliminary results on the charge asymmetry for these decays have been given by Thresher [12]. For the 3π decays a value of 0.0094 ± 0.0081 was obtained and for the $\pi\pi\gamma$ decays a value of 0.11 ± 0.10, both clearly being consistent with zero. The rate asymmetry for the 3π decays was found to be 0.0025 ± 0.0053 which is again consistent with zero, and also with the value of -0.011 ± 0.018 of Herzo et al. [13]. There is obviously no evidence for CP violation in either of the decay modes.

Direct tests for an isotensor component in the electromagnetic current have been proposed in nuclear physics and a summary of the evidence has been given by Blin-Stoyle [14]. The first test that was suggested is to look for deviations from the quadratic mass formula

$$M = M_0(1 + a I_z + b I_z^2)$$

which should hold for members of a nuclear isospin multiplet assuming only isoscalar and isovector properties for the electromagnetic interaction. The second test is to look for $\Delta I = 2$ γ-transitions, and the third test, suggested initially by Blin-Stoyle [14] is to compare the decay widths for $I = \frac{3}{2} \to I = \frac{1}{2}$ transitions in mirror nuclei ($I_z = \pm\frac{1}{2}$).

The difficulty with nuclear physics tests is that since the electromagnetic interaction of individual nucleons can only contain $|\Delta I| = 0, 1$ parts, the effects being investigated are strongly suppressed. Indeed, any nuclear electromagnetic interaction which can be well described by the sum of individual nucleon contributions without explicitly taking account of mesonic and multinucleon effects necessarily leads to the $|\Delta I| = 0, 1$ rule.

Thus with the first test, any two nucleon interaction arising from an isotensor component will be short range and, at best, can contribute in first order to quadratic terms in the mass formula [15]. For the test to be effective, either theoretical and experimental values of a and b have to be compared, involving the uncertainties about nuclear wavefunctions and the charge dependence of nuclear forces, or the term arising from a three-body charge-dependent force has to be found. Results on the latter for the mass 9 quartet ^9C, ^9B, ^9Be and ^9Li have been quoted by Adler [8]. The coefficient c of the possible cubic term was found to have an upper bound of $0.04\,b$, which is consistent with the magnitude $c \sim \alpha b \sim 0.01\,b$ expected from conventional fourth-order electromagnetic effects. This, of course, does not eliminate an isotensor component, because of the very high suppression which one would expect.

The difficulties in the second test are twofold. Firstly there are not many examples of established states in the same nucleus differing by $\Delta I = 2$ and secondly they are difficult to observe, since they lie above so many intermediate states that direct transitions to the $I = 0$ ground state are relatively unlikely and transitions to excited $I = 0$ states are extremely difficult to identify. Nevertheless, Snover et al. [16] have set a limit on the γ width for the decay from the $0^+, I = 2$ state at 15.2 MeV to the $2^+, I = 0$ state at 1.78 MeV in ^{28}Si. The upper limit obtained was 0.03 Γ_γ (Γ_γ being the total γ width of the 15.2 MeV state), implying that the $I = 2 \gamma$ decay amplitude is less than about 20% of the $I = 1$ amplitude.

For the third test, if A_1 and A_2 are the amplitudes for the isovector and isotensor transitions respectively, then Blin-Stoyle [14] showed that for the mirror nuclei ^{13}C and ^{13}N, a value of $A_2/A_1 = -0.026 \pm 0.116$ which is clearly consistent with zero but sets an upper limit of about 15% on the relative isotensor amplitude. Consideration of more complicated decays in the same two nuclei, allowed the upper limit to be reduced to the order of 10%.

2. Pion Photoproduction and π^- Radiative Capture

In principle, to test for T-violation in this case is straightforward. One simply invokes detailed balance and compares, for example, the reaction $\gamma + n \rightarrow \pi^- + p$ with the inverse one $\pi^- + p \rightarrow \gamma + n$. Any difference observed can be due only to T-violation. In practice, however, it is much more complicated. The photoproduction reaction must necessarily be done on deuterium, so that one is immediately involved with the whole problem of deuterium corrections and the capture process has to be observed against the large background coming from the charge exchange reaction $\pi^- + p \rightarrow \pi^0 + n$.

For the differential cross sections for π^- photoproduction, there are two extensive bubble chamber experiments [17, 18] using the spectator model to extract the free neutron cross-sections from deuterium, and three differential cross-section points at 275 MeV by Garelick and Cooperstein [19]. This latter experiment detected both the π^- and proton in coincidence and performed a modified Chew-Low extrapolation to obtain the cross-section. There is a very important experiment at 180° by Fujii et al. [20]. In this experiment, both π^+ and π^- production on deuterium was measured, and additionally π^+ on protons. The measurements allow a definitive check of the deuterium corrections which are found to be small, as would be expected for backward production, and enable one to have considerable faith in the π^- results. Below resonance, there are a number of π^-/π^+ ratio measurements in deuterium, which have been discussed in some detail by Hogg [21].

For π^- radiative capture, there is an excitation curve at approximately 30° c.m. from Favier *et al.* [22], subsequently modified by Schinzel [23] and two angular distributions at equivalent photon laboratory energies of 350 MeV, 480 MeV and 520 MeV by Berardo *et al.* [24]. A comparison of these three latter with the data in the direct reaction is given in Fig. 1a—c respectively. There is no obvious discrepancy at the two higher energies;

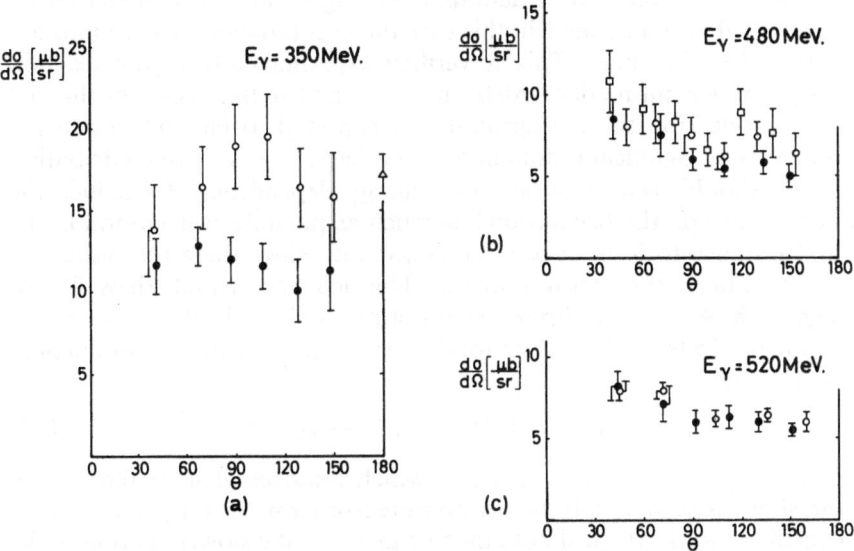

Fig. 1. A comparison of π^- photoproduction data (open circles, squares and triangles) with π^- radiative capture data (solid circles). The data are from the ABBHHM collaboration [17] (open circles), the PFRN collaboration [18] (open squares), Fujii *et al.* [20] (open triangle) and Berardo *et al.* [24] (solid circles)

there is possibly one at the lower, the deuteron measurements for $\theta \geq 90°$ being, on the average, some 30% larger than the $\pi^- + p \to \gamma + n$ values. This becomes more quantitative if one considers the 180° values alone. Any reasonable Legendre polynomial fit to the radiative capture data yields a backward cross-section of $10 \pm 2\,\mu b/sr$ while the result from Fujii *et al.* [20] is $17 \pm 1\,\mu b/sr$. If both these experiments are correct then the existence of some T-violation is established, and this interpretation has indeed been made [25]. The agreement of the radiative capture and photoproduction data at 480 MeV and 520 MeV is readily explicable in terms of our conjecture that any T-violating effects should be primarily connected with the resonance.

The observation of an isotensor component in the electromagnetic current is not quite so clear cut. The objective is to compare the radiative widths of the Δ^+, Δ^0 charge states of the resonance. If the parameter x defined by

$$\Gamma(\Delta^0 \to n\gamma) = \Gamma(\Delta^+ \to p\gamma)(1+x)^2 \qquad (2.1)$$

is non-zero, then an isotensor term must be present. The two pairs of reactions which present themselves for comparison are $\gamma + n \to \pi^0 + n$ and $\gamma + p \to \pi^0 + p$ on the one hand and $\gamma + n \to \pi^- + p$ and $\gamma + p \to \pi^+ + n$ on the other hand. The reactions involving π^0 are to be preferred on theoretical grounds, since in this case the nonresonant backgrounds are expected to be small. This is verified experimentally in the reaction $\gamma + p \to \pi^0 + p$ but no data exists on $\gamma + n \to \pi^0 + n$. In the case of charged pion production the backgrounds are expected to be, and indeed are found to be significant, dominated by a large, real s-wave contribution which should have a rather slow energy dependence. To reduce the dependence on the background as much as possible, it is natural in the first instance to look at total cross-sections, since there the resonance s-wave interference term vanishes. This led Sanda and Shaw [5] to suggest looking for a dip (or peak) of approximately the width of the resonance in the difference of total cross-sections σ_T for $\pi^\pm$ production, i.e. to study

$$\Delta'(W) = (k/q)\{\sigma_T(\pi^-) - \sigma_T(\pi^+)\} \qquad (2.2)$$

where k/q is a phase space factor which removes at least the s-wave threshold dependence. If there is no isotensor term, the resonance cannot contribute to Δ' which should then be given by the slowly varying background. The data, including total cross-section measurements of $\gamma + n \to \pi^- + p$ by White *et al.* [26] are shown in Fig. 2. The total cross-section for $\gamma + p \to \pi^+ + n$ has been taken from detailed multipole analyses of the proton data by Noelle, Pfeil and Schwela [27] and Berends and Weaver [28]. Despite the considerable experimental uncertainty there is clear evidence for a dip of the kind required by an isotensor current.

Thus the experimental data are indicating the possibility of a T-violating component and an isotensor component in the electromagnetic current. A discussion of T-violation in pion photoproduction was first given by Christ and Lee [29] and we have traced the development of the isotensor argument in pion photoproduction through the work of Shaw [4] and Sanda and Shaw [5]. The two aspects were combined by Sanda and Shaw [30] who postulated an isotensor T-violating electromagnetic current and showed that such a postulate was able to provide a rather clear qualitative understanding of the various T(or CP) violating effects in electromagnetic interactions. In particular it explained simultaneously the structure observed in $\Delta'(W)$, Eq. (22), and the discrepancy between π^- photoproduction and π^- radiative capture. Finally a complete quantitative account of all the data in the first resonance region has been given by Donnachie and Shaw [31] in terms of a detailed development of the model of Sanda and Shaw [30], incorporating the possibility of T-violation in both the isovector and isotensor amplitudes.

Considering the process

$$\gamma(K) + N_1(P_1) \rightleftharpoons \pi_\alpha(Q) + N_2(P_2) \tag{2.2}$$

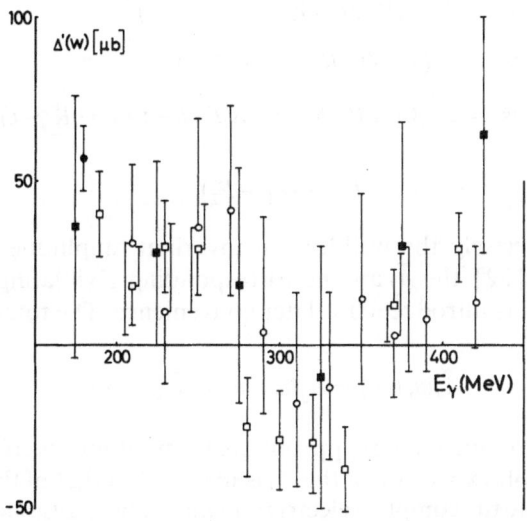

Fig. 2. The difference between the π^- and π^+ photoproduction total cross sections. The data are from the ABBHHM collaboration [17] (open circles), the PFRN collaboration [18] (open squares), White *et al.* [26] (solid squares) and π^-/π^+ ratios (solid circle)

in the presence of C-conserving and C-violating currents, the transition matrix can be written as

$$\langle \pi N_2 | T | \gamma N_1 \rangle$$
$$= -i(2\pi)^{5/2}(2k)^{-1/2}\delta^4(P_2 + Q - P_1 - K)\varepsilon_{\mu \text{out}} \langle \pi N_2 | \{J_\mu^{(0)} + K_\mu^{(0)}\} | N_1 \rangle_{\text{in}}$$
$$= -i(2\pi)^{-2}\delta^4(P_2 + Q - P_1 - K)\tfrac{1}{2}m_N(k\omega_q E_1 E_2)^{-1/2} \langle \pi N_2 | T_e + T_0 | \gamma N_1 \rangle \tag{2.3}$$

where the T_e, T_0 terms arise from the C-conserving and C-violating currents repectively. For the photoproduction process $\gamma N \to \pi N$ the relevant amplitude is $\langle \pi N_2 | T_e + T_0 | \gamma N_1 \rangle$ and for the time reversed radiative capture process $\pi N \to \gamma N$ the relevant amplitude is

$$\langle \pi N_2 | T_e - T_0 | \gamma N_1 \rangle .$$

The invariant amplitude decomposition for the C-violating terms is precisely the same as that for the C-conserving terms, and we can write

$$\langle \pi N_2 | T | \gamma N_1 \rangle = \sum_{j=1}^{4} \{A_j(s,t,u) + i\bar{A}_j(s,t,u)\} \, \bar{u}(P_2) \, M_j u(P_1) \quad (2.4)$$

where

$$M_1 = i\gamma_5 \gamma \cdot \varepsilon\gamma \cdot K$$

$$M_2 = 2i\gamma_5 (P \cdot \varepsilon Q \cdot K - P \cdot K Q \cdot \varepsilon)$$

$$M_3 = \gamma_5 (\gamma \cdot \varepsilon Q \cdot K - \gamma \cdot K Q \cdot \varepsilon)$$

$$M_4 = 2\gamma_5 (\gamma \cdot \varepsilon P \cdot K - \gamma \cdot K P \cdot \varepsilon - i m_N \gamma \cdot K \gamma \cdot \varepsilon)$$

$$(2.5)$$

with

$$P = \tfrac{1}{2}(P_1 + P_2). \quad (2.6)$$

The A_j are precisely the well-known invariant amplitudes introduced by Chew *et al.* [32], the $\bar{A}_j$ are the corresponding T-violating amplitudes and the factor i is introduced for later convenience. The total amplitudes are obviously

$$\tilde{A}_j(s,t,u) = A_j(s,t,u) \pm \bar{A}_j(s,t,u) \quad (2.7)$$

where the plus (minus) sign is appropriate for photoproduction (capture).

The multipole expansion of the $\bar{A}_j$ is identical to that of the A_j [32] so that we can write the complete electric and magnetic multipole amplitudes $\tilde{E}_{l\pm}, \tilde{M}_{l\pm}$ leading to final πN states of angular momentum $J = l \pm \tfrac{1}{2}$ as

$$\tilde{E}_{l\pm} = E_{l\pm} + i\bar{E}_{l\pm}, \quad \tilde{M}_{l\pm} = M_{l\pm} + i\bar{M}_{l\pm}$$

$$\tilde{E}_{l\pm} = E_{l\pm} - i\bar{E}_{l\pm}, \quad \tilde{M}_{l\pm} = M_{l\pm} - i\bar{M}_{l\pm}$$

$$(2.8)$$

for photoproduction and capture respectively.

In addition to the usual isospin amplitudes for photoproduction [32] i.e. an isoscalar amplitude $A^{(0)}$ leading to the $I = \tfrac{1}{2}$ final πN state and isovector amplitudes $A^{(1)}, A^{(3)}$ leading to the $I = \tfrac{1}{2}, \tfrac{3}{2}$ final πN states respectively, we have also the isotensor amplitude $A^{(3)}$ leading to the $I = \tfrac{3}{2}$ final πN state. The amplitudes for observable processes are

$$A(\gamma p \to \pi^+ n) = \sqrt{2/3}(3A^{(0)} + A^{(1)} + \sqrt{3/5}\,A^{(2)} - A^{(3)})$$

$$A(\gamma n \to \pi^- p) = \sqrt{2/3}(3A^{(0)} - A^{(1)} + \sqrt{3/5}\,A^{(2)} + A^{(3)})$$

$$A(\gamma p \to \pi^0 p) = \tfrac{1}{3}(3A^{(0)} + A^{(1)} - 2\sqrt{3/5}\,A^{(2)} + 2A^{(3)})$$

$$A(\gamma n \to \pi^0 n) = \tfrac{1}{3}(-3A^{(0)} + A^{(1)} + 2\sqrt{3/5}\,A^{(2)} + 2A^{(3)}).$$

$$(2.9)$$

It is convenient to introduce amplitudes leading to the $I = \frac{1}{2}, \frac{3}{2} \pi N$ final states on protons and neutrons, namely

$$
\begin{aligned}
_pA^{(3)} &= \sqrt{2/3}\,\{A^{(3)} - \sqrt{3/5}\,A^{(2)}\} \\
_nA^{(3)} &= \sqrt{2/3}\,\{A^{(3)} + \sqrt{3/5}\,A^{(2)}\} \\
_pA^{(1)} &= -\sqrt{1/3}\,\{A^{(1)} + 3\,A^{(0)}\} \\
_nA^{(1)} &= \sqrt{1/3}\,\{A^{(1)} - 3\,A^{(0)}\}\,.
\end{aligned}
\tag{2.10}
$$

For the conventional $C = -1$ amplitudes, the content of Watson's theorem [33] is well known, namely that below the $\pi\pi N$ threshold unitarity and T-invariance lead to the result that the multipole amplitudes $E_{l\pm}^{(I)}$, $M_{l\pm}^{(I)}$ have the same phase (modulo π) as the πN scattering amplitude $f_{l\pm}^{(I)}$ leading to the same final state. For an amplitude odd under T similar arguments lead to an amplitude $90°$ out of phase with $f_{l\pm}^{(I)}$. In our definition of the T-violating multipoles $\bar{E}_{l\pm}^{(I)}$, $\bar{M}_{l\pm}^{(I)}$ (Eq. (28)) this factor i has been removed so that the phase of these amplitudes is also given by the phase shift (modulo π) below the $\pi\pi N$ threshold.

Thus the complete multipole amplitudes can be written

$$
\begin{aligned}
\tilde{E}_{l\pm}^{(I)} &= E_{l\pm}^{(I)} \pm i\bar{E}_{l\pm}^{(I)} = (\{E_{l\pm}^{(I)}\} \pm i\{\bar{E}_{l\pm}^{(I)}\})\exp i\delta_{l\pm}^{(I)} \\
\tilde{M}_{l\pm}^{(I)} &= M_{l\pm}^{(I)} \pm i\bar{M}_{l\pm}^{(I)} = (\{M_{l\pm}^{(I)}\} \pm i\{\bar{M}_{l\pm}^{(I)}\})\exp i\delta_{l\pm}^{(I)}
\end{aligned}
\tag{2.11}
$$

where the plus (minus) sign is appropriate for photoproduction (capture) and { } means $\pm$ the modulus.

It can also be written

$$
\begin{aligned}
\tilde{E}_{l\pm}^{(I)} &= \{\tilde{E}_{l\pm}^{(I)}\}\,(\exp i\delta_{l\pm}^{(I)})\,(\exp \pm i\phi_{l\pm}^{(I)}) \\
\tilde{M}_{l\pm}^{(I)} &= \{\tilde{M}_{l\pm}^{(I)}\}\,(\exp i\delta_{l\pm}^{(I)})\,(\exp \pm i\psi_{l\pm}^{(I)})
\end{aligned}
\tag{2.12}
$$

and while this is sometimes convenient as a parametrization it is not convenient for theoretical discussion since A and $\bar{A}$ have quite different analytic structures and it is these amplitudes, but not $\tilde{A}$ that have simple crossing properties. Explicitly, the C-violating amplitudes have opposite crossing properties to the C-conserving amplitudes, and the isotensor amplitude has the same crossing properties as the isoscalar amplitudes. The crossing properties of the conventional amplitudes are well known and well documented.

Ever since the original work of Chew et al. [32], the conventional description of pion photoproduction has been in terms of fixed t dispersion relations, which have had remarkable success. A review of the formalism and a discussion of the development of the subject has been given by Donnachie [34]. The model of Donnachie and Shaw [31] is based on these dispersion relations.

The singularities of the non-exotic amplitudes are assumed to be dominated by the diagrams of Fig. 3. On the other hand, since the $I = 2$ and C-violating terms cannot couple to the Born term poles, Fig. 3 a–c, but can only couple in the $\Delta N \gamma$ vertex only diagrams 3 d and 3 e will contribute in this case. The first of these two will dominate, being direct excitation of the resonance, and the second will only serve to impart small corrections via crossing. Given the present state of the neutron

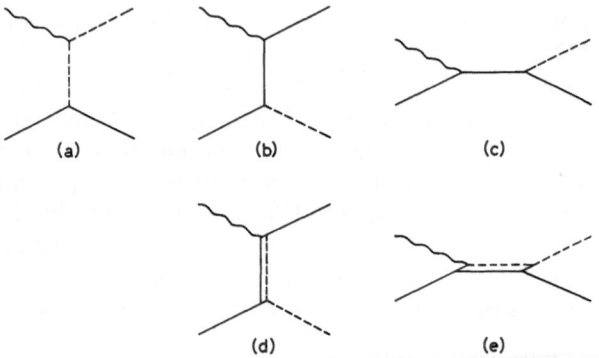

Fig. 3 a–e. Born terms and contributions to the low energy absorptive parts for pion photoproduction

data, these small corrections can be neglected. Donnachie and Shaw [31] made the additional assumption that the resonance excitation is entirely magnetic dipole. It is known that the electric quadrupole amplitude on protons, $_pE_{1+}^{(3)}$ is much smaller at resonance than the magnetic dipole amplitude $_pM_{1+}^{(3)}$ (about -4%) and the assumption being made is that this is true on neutrons also. Note that in the absence of isotensor terms $_pE_{1+}^{(3)}$ and $_nE_{1+}^{(3)}$ are identically equal so that there is no assumption at all in this case.

It was assumed that the resonance shape was the same for both the isovector and isotensor amplitudes, namely

$$M(w) = A k q^{-2} (\exp i \delta_{33}) \sin(\delta_{33} + x_1 q^3/k^3) \qquad (2.13)$$

where A is a (suitably chosen) constant and x_1 a parameter. This reduces to the usual resonance form if the phase x_1 is zero. T-violating phases were allowed in both the isovector and isotensor parts, the multipoles being given by

$$\begin{aligned}
\tilde{M}_{1+}^{(3)} &= M(W) x_2 \exp i x_4 \\
\tilde{M}_{1+}^{(2)} &= M(W) x_3 \exp i x_5
\end{aligned} \qquad (2.14)$$

where $x_2, \ldots, x_5$ are again parameters, the T-violating phases x_4, x_5 changing sign under T-reversal. Strictly speaking, x_4 cannot be exactly constant since

$$M_{1+}^{(3)} = M(W) x_2 \cos x_4$$

must contain the Born term poles, whereas

$$\bar{M}_{1+}^{(3)} = M(W) x_2 \sin x_4$$

must not. However to the extent that the amplitude is dominated by the second sheet resonance poles, x_4 is constant, so that the approximation should be a good one.

It is also convenient to define corresponding parameters for $_p\bar{M}_{1+}^{(3)}$, $_n\bar{M}_{1+}^{(3)}$, i.e.

$$_p\tilde{M}_{1+}^{(3)} = \sqrt{2/3} M(W) \{x_2 \exp i x_4 - \sqrt{3/5} x_3 \exp i x_5\} = \sqrt{3/3} M(W) x_p \exp i \phi_p$$
$$_n\tilde{M}_{1+}^{(3)} = \sqrt{2/3} M(W) \{x_2 \exp i x_4 + \sqrt{3/5} x_3 \exp i x_5\} = \sqrt{2/3} M(W) x_n \exp i \phi_n.$$

$$(2.15)$$

With two exceptions, all other multipoles up to and including transitions to f-waves were taken from conventional dispersion relation calculations, using the numerical values given by Berends, Donnachie and Weaver [35]. All transitions to higher partial waves were included in the Born approximation. The exceptions were the magnetic dipole transitions $M_{1-}^{(0)}$, $M_{1-}^{(1)}$ which cannot be calculated theoretically because of the nearby presence of the $P_{11}(1450)$ resonance. A simple rescattering correction to the theoretical contributions to these multipoles was made, involving one additional parameter in each case. Thus the model of Donnachie and Shaw [31] attempts a description of these processes in terms of only seven parameters, the four interesting ones being the isovector, isotensor resonance coupling strengths x_2, x_3 and the isovector, isotensor T-violating phases x_4, x_5.

An excellent fit to the data was obtained, and the model could not account for the data without the presence of both C-violating and isotensor terms. The stability of the parameters against detailed variations in the model and against possible ambiguities in the proton data was investigated at length and although there was some variation in the values of the parameters, both the ratio of isovector to isotensor resonance excitation and the T-violating phase on neutrons were found to be essentially invariant against all the modifications attempted. It is important to note that conventional fits to the data look rather implausible in terms of fixed $- t$ dispersion relations, and the inclusion of the corrections to the latter from higher resonances are demonstrably not sufficient. This has been looked at in particular by Devenish, Lyth and Rankin [36], using the results of their own extensive phenomenological analysis of the second, third and fourth resonance regions [37], and they conclude that all such effects are small compared to the errors on the π^- data and do not affect the general conclusions about an isotensor, or C-violating, term.

In discussing the results, it is convenient to introduce the subsidiary parameters

$$x = (x_n - x_p)/x_p, \qquad t = x_2/x_3 . \tag{2.16}$$

In the absence of an isotensor term $x_p - x_n = x = t = 0$.

Obviously the results are only reliable to the extent that they are not sensitive to the details of the model. How are the parameters determined by the data? The resonance strengths x_p, x_n will be fixed essentially by the total cross-sections, or general magnitude of the data. The T-violating phase ϕ_n is determined by the fact that it changes sign between the reactions $\gamma n \rightarrow \pi^- p, \pi^- p \rightarrow \gamma n$ and its existence is simply a matter of the data difference which has been observed. The magnitude of the effect is governed by the interference between the T-violating part of the resonant amplitude and the background, and since this is largely given by the Born term the magnitude of ϕ_n will be reasonably independent of the details of the model. On the other hand, since there is no inverse data on proton reactions the existence of ϕ_p cannot be independently established and any value obtained well be model dependent.

These general remarks were borne out by the analysis of Donnachie and Shaw [31] and expressing their results in terms of the parameters x, t and ϕ_n

$$-0.28 \geqq x \geqq -0.36$$

$$-0.23 \geqq t \quad \geqq -0.31 \tag{2.17}$$

$$-7.9° \geqq \phi_n \geqq -11.4° .$$

The T-violating phase on protons ϕ_p and hence the isovector and isotensor T-violating phases ϕ_1 and ϕ_2 are essentially undetermined. At present, they can only be determined by making additional ad hoc assumptions. For example, if we assume pure isotensor T-violation, as advocated by Sanda and Shaw [30] on the strength of the arguments outlined in the Introduction, then taking $t = -0.28$, $\phi_n = -11°$ as representative values

Pure isotensor:

$$\begin{aligned} \phi_1 &= 0° & \phi_2 &= 51° \\ \phi_p &= 7.6° & \phi_n &= -11° \\ x &= -0.31 . \end{aligned} \tag{2.18}$$

On the other hand, if we assume the T-violation to be purely isovector, then with the same values of t and ϕ_n

Pure isovector:

$$\phi_1 = -8.6° \qquad \phi_2 = 0°$$
$$\phi_p = -7.1° \qquad \phi_n = -11° \qquad (2.19)$$
$$x = -0.35.$$

3. Deuteron Photodisintegration and Neutron-Proton Radiative Capture

Another test of T-violation through detailed balance is by a comparison of the reactions $\gamma + d \to n + p$ and $n + p \to \gamma + d$. The complication of the deuterium corrections, which are a serious problem in the photoproduction test, is no longer present but the capture process has again to be viewed against a large background, this time coming from $n + p \to \pi^0 + d$. However the theoretical interpretation is not as simple as in the photoproduction case. Any difference observed is undeniably due to T-violation, but the interpretation, in terms of a phase for example, can only be done in a very model dependent way. Further, since the deuteron is an isoscalar, only isovector T-violation is being investigated. An isotensor T-violating term would give no effect.

The relevant energy region is about 290 MeV photon laboratory energy or equivalently, 590 MeV neutron laboratory kinetic energy. In this energy region there is a well known peak in the total cross-section for deuteron photodisintegration, Fig. 4, the data being taken from Keck and Tollestrup [38] and Whalin, Schriever and Hanson [39]. The peak is undoubtedly associated with the $P_{33}(1236)$ and a possible mechanism is shown in Fig. 5.

The existence of the Δ in the intermediate state and consequentally the $\Delta N \gamma$ vertex is the reason why this process is of prime interest in studying T-violation. This was first realized by Barshay [40], who estimated the possible magnitude of the effect, using a model due to Austern [41] and inserting a T-violating phase in the $\Delta N \gamma$ vertex. The model is simply to evaluate specifically the time-ordered Feynman graph of Fig. 5. The form of the resultant matrix element M is a kinematical factor multiplying essentially the resonance propagator and a projection operator for the $^1D_2$ state with $I = 1$ of the final two-nucleon system, this being the dominant transition induced by the mechanism. The kinematical factor involves an integral over the deuteron internal momentum which requires a cut-off. This cut-off is the only unknown parameter in the model and can be determined from the experimental

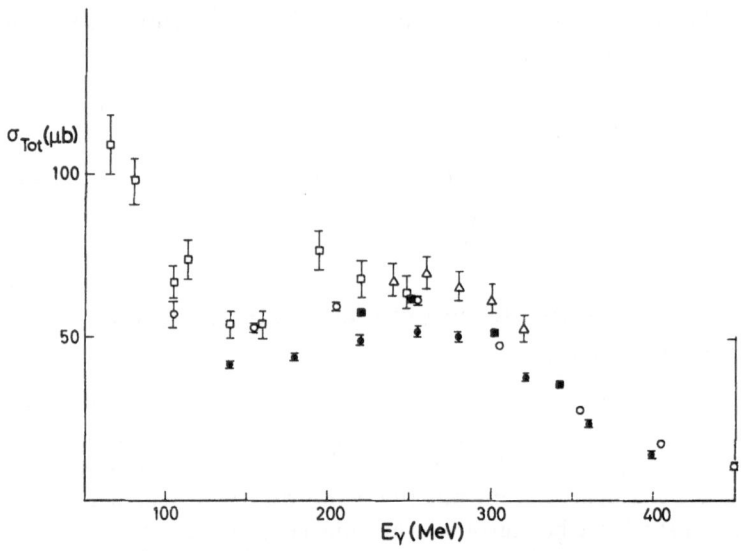

Fig. 4. The total cross section for deuteron photodisintegration. The data are from Keck *et al.* [38] (open circles), Whalin *et al.* [39] (open squares), Sober *et al.* [42] (open triangles), Buon *et al.* [43] (solid circles) and Anderson *et al.* [44] (solid squares)

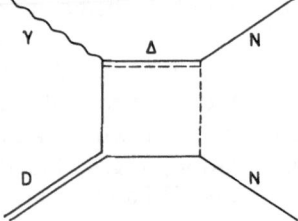

Fig. 5. Typical mechanism for Δ excitation in deuteron photodisintegration

total cross-section by equating the observed height above background of the bump with the value calculated from the matrix element.

Insertion of the *T*-violating phase ϕ into the $\gamma N \Delta$ vertex simply converts the matrix element to $M \exp i\phi$ for the photodisintegration process and to $M \exp - i\phi$ for the inverse neutron-proton capture process. The total cross-sections for the two processes are thus equal and the effect of *T*-violation will show up only in a difference in shape of the angular distributions. This will involve an interference of the resonance amplitude with the background terms, and it is thus necessary to be able to calculate these as well to allow any determination of the phase ϕ should an effect be observed.

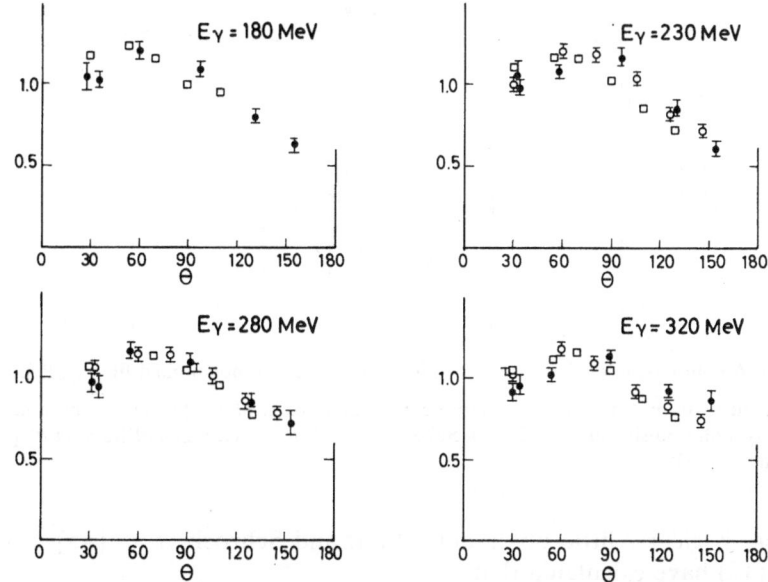

Fig. 6. A comparison of the neutron-proton radiative capture data of Schrock *et al.* [45] (solid circles) with deuteron photodisintegration data of Sober *et al.* [42] (open circles) and Buon *et al.* [43] (open squares). To facilitate comparison of the shape, the data are arbitrarily normalized to a total cross section of 4π

The dominant background transitions are electric dipole to a $^3P_0$ final state and magnetic dipole to a $^1S_0$ final state, contributing about $45\,\mu$b and $3\,\mu$b respectively to the total cross-section, compared to $27\,\mu$b from the resonant term. In the differential cross-section only the singlet states will interfere and because of the small magnitude of the $^1S_0$ transition the T-violating effects, if present, will be strongly suppressed. If the differential cross-section is expanded in a Legendre polynomial power series

$$\mathrm{d}\sigma/\mathrm{d}\Omega = A_0 + A_1 P_1 (\cos\vartheta) + A_2 P_2 (\cos\vartheta) \qquad (3.1)$$

then the $^1S_0$, $^1D_2$ interference terms will occur in A_0 and A_2. Taking the background transitions to be purely real (i.e. assuming that the real parts of the $n-p$ $^1S_0$ and $^3P_0$ phase shifts are small at the energies considered), Barshay [40] showed that the effect could be expressed as

$$\sin\phi \approx 3\{(A_2/A_0)(n+p\rightarrow\gamma+d) - (A_2/A_0)(\gamma+d\rightarrow n+p)\} . \qquad (3.2)$$

Recent photodisintegration data (Sober *et al.* [42], Buon *et al.* [43] and Anderson *et al.* [44]) are compared with recent neutron proton radiative capture data (Schrock *et al.* [45]) in Fig. 6 and ratio $-(A_2/A_0)$ in Fig. 7.

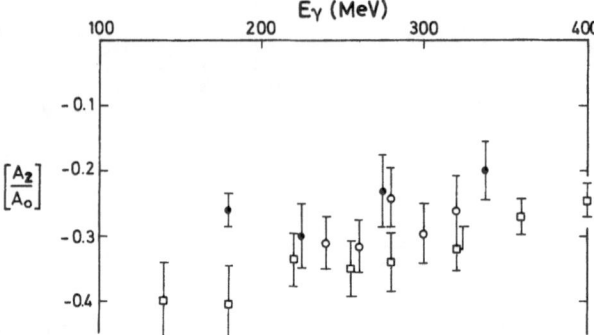

Fig. 7. A comparison of the ratio A_2/A_0 from the Legendre polynomial fit $\sum_n A_n P_n(\cos \vartheta)$ to the neutron proton radiative capture data of Schrock et al. [43] (solid circles) and the deuteron photodisintegration data of Sober et al. [42] (open circles) and Buon et al. [43] (open squares)

There is clearly little observable effect, and Schrock et al. [45], using Eq. (3.2) have calculated that

$$\phi = 4° \pm 10° \tag{3.3}$$

at the resonance peak, i.e. is consistent with time-reversal invariance. However A is also consistent with the pure isovector T-violating phase found by Donnachie and Shaw [31] in pion photoproduction, i.e. the deuteron photodisintegration/neutron-proton radiative capture cannot tell us the answer one way or the other.

4. Inelastic Electron Scattering off a Polarized Target

In inelastic electron scattering off a polarized target, a difference in cross-sections (i.e. an asymmetry) when the target is polarized in opposite directions along the normal to the scattering plane is indicative of T-violation. The only possible alternative explanation would be from higher order electromagnetic effects which one would not expect to be able to observe, and which can, in any case, be checked explicitly by using both electron and positron beams.

The relevant theoretical formalism has been developed by Christ and Lee [46]. Inelastic electron scattering is described by three form factors, $F_\pm$ corresponding to the two transverse polarization states of the virtual photon and F_z corresponding to the longitudinal (or alternatively scalar) polarization state of the virtual photon.

Since the final hadronic state is not studied, it is convenient to work entirely in the laboratory frame and we shall use the following notation.

$(\omega, \boldsymbol{q}), (\omega', \boldsymbol{q}')$ the four-momenta of the initial and final electron respectively,

ψ the scattering angle of the electron,

$(E, \boldsymbol{P})$ the four-momentum of the final hadron state Γ,

W_Γ the effective mass of the final hadron state Γ,

(the four-momentum transfer)$^2 = (\boldsymbol{q} - \boldsymbol{q}')^2 - (\omega - \omega')^2$
$$= 4\omega\omega'\sin^2(\tfrac{1}{2}\psi),$$

v the energy of the virtual photon $= \omega - \omega' = q_0$,

s_i the spin of the target proton,

$\hat{n}$ the normal to the scattering plane $= (\boldsymbol{q} \times \boldsymbol{q}')/|\boldsymbol{q} \times \boldsymbol{q}'|$.

The double differential cross section in the usual notation is

$$\frac{\mathrm{d}^2\sigma}{\mathrm{d}\Omega\mathrm{d}\omega'} = \Gamma_t\left[\sigma_T + \varepsilon\sigma_L + \sqrt{2\varepsilon(\varepsilon+1)}\,\frac{k_0}{k}\,\sigma_I(s_i \cdot \hat{n})\right] \tag{4.1}$$

where σ_T, σ_L and σ_I are the cross-sections arising from the transverse, longitudinal and transverse-longitudinal interference parts of the amplitude

$$\Gamma_t = \frac{\alpha}{4\pi^2}\,\frac{\left(v - \dfrac{q^2}{2m_N}\right)2\omega'}{q^2(1-\varepsilon)\,\omega} \tag{4.2}$$

and

$$\varepsilon = \frac{1}{1 + 2(1 + v^2/q^2)\tan^2(\psi/2)}. \tag{4.3}$$

On terms of the form factors and the quantities W_1, W_2 and W_3 introduced by Christ and Lee [46]

$$\begin{aligned}
\sigma_T &= \frac{8\pi^2\alpha m_N}{W_\Gamma^2 - m_N^2}\sum_\Gamma\{|F_+|^2 + |F_-|^2\} = \frac{8\pi^2\alpha m_N}{W_\Gamma^2 - m_N^2}W_1 \\[2mm]
\sigma_L &= \frac{8\pi^2\alpha m_N}{W_\Gamma^2 - m_N^2}\frac{K^2}{v^2}\sum_\Gamma|F_z|^2 = \frac{8\pi^2\alpha m_N}{W_\Gamma^2 - m_N^2}\left\{\left(1 + \frac{v^2}{K^2}\right)W_2 - W_1\right\} \\[2mm]
\sigma_I &= \frac{8\pi^2\alpha m_N}{W_\Gamma^2 - m_N^2}2\theta\sum_\Gamma\mathrm{Im}(F_z^* F_-) = \frac{8\pi^2\alpha(1 + v^2/K^2)}{W_\Gamma m_N(W_\Gamma^2 - m_N^2)}v^2 W_3
\end{aligned} \tag{4.4}$$

where η is a phase factor depending on both the spin and parity of the final state Γ.

It is the third term, σ_I (or W_3) which produces a spin-dependent cross-section, i.e. an asymmetry. If there were no T-violation then all the form-factors would have the same phase ensuring that $\sigma_I \equiv W_3 \equiv 0$. Defining the asymmetry parameter as

$$A = \frac{\sigma_\uparrow - \sigma_\downarrow}{\sigma_\uparrow + \sigma_\downarrow} \tag{4.5}$$

and using Eqs. (4.2), (4.4) we get

$$|A| = \frac{\sqrt{2\varepsilon(\varepsilon+1)}\sin|\phi|\sqrt{\sigma_L/\sigma_T}}{1+\varepsilon(\sigma_L/\sigma_T)} \frac{|F_-|}{[|F_+|^2+|F_-|^2]^{\frac{1}{2}}} \tag{4.6}$$

where we have taken only one term in the summation in Eq. (4.4) and ϕ is the phase between F_- and F_z.

With our present knowledge of inelastic electron scattering we can proceed beyond this point only by means of rather rough estimates. As a first approximation, it should be noted that in the kinematic region experimentally accessible $\varepsilon \cong 1$. Secondly we know from a study of the reaction $ep \rightarrow ep\pi^0$ in the first resonance region (Albrecht *et al.* [47], Hellings *et al.* [48], a review of which has been given recently by Gayler [48]) that, as in photoproduction, the magnetic dipole term dominates the cross-section. If this were the only term present, then

$$|F_-|/\{|F_+|^2+|F_-|^2\}^{\frac{1}{2}} = \tfrac{1}{2}$$

and so with these approximations we finally have

$$|A| = \frac{\sqrt{\sigma_L/\sigma_T}}{1+\sigma_L/\sigma_T}\sin|\phi| . \tag{4.7}$$

Further, if we assume (as we did in photoproduction) that the *T*-violation is associated entirely with the magnetic dipole term, then the phase ϕ introduced here is the same as the *T*-violating phases introduced earlier.

Experimentally, one is faced with a problem. Since the effect depends on the longitudinal excitation of the resonance, reasonably large values of K^2 must be used to ensure that this is observable but, on the other hand, resonance excitation decreases roughly as a dipole term (in the amplitude) with increasing K^2 and the resonance very rapidly disappears. Further σ_L/σ_T never becomes very large anyway, and the method is not as sensitive as had been initially hoped.

Attempts to measure this asymmetry have been made by Chen *et al.* [50] and by Rock *et al.* [51]. The only apparent asymmetry observed was by the latter group at $K^2 = 0.6 \,(\text{GeV}/c)^2$, obtaining an average asymmetry over three points in the region of the $P_{33}(1236)$ resonance of $(4.5 \pm 1.4)\%$. This 3.2 standard deviation asymmetry has a 0.14% chance of occurring in this particular spot but, as the authors themselves point out, there is a 15% statistical chance for A to appear somewhere in the data which is equivalent to less than 2.0 standard deviations.

If one takes this effect for real, and takes the recently obtained values of $|\sigma_L/\sigma_T| \simeq 0.1$ [47–49] then A corresponds to

$$|\phi| \simeq (20 \pm 6)^\circ . \tag{4.8}$$

5. The Neutron Electric Dipole Moment

A non-zero value of the neutron electric dipole moment can arise only from a combination of both P-violation and T-violation. The standard theoretical approach is that of sidewise dispersion relations developed by Bincer [52], Drell and Pagels [53], Barton and White [54] and finally by Broadhurst [55].

The off-shell vertex function and its absorptive part are defined by

$$\Gamma_\mu (p_2, p_1) = i(p_{20}/m)^{\frac{1}{2}} \int d^4x \exp - i p_1 x [\langle p_2 | \theta(-x_0) [j_\mu(0), \bar{\psi}(x)] | 0 \rangle D_x]$$

(5.1)

$$A_\mu(p_2, p_1) = \tfrac{1}{2}(p_{20}/m)^{\frac{1}{2}} \int d^4x \exp - i p_1 x \langle p_2 | [j_\mu(0), \bar{\chi}(x)] | 0 \rangle \quad (5.2)$$

where $p_1^2 = W^2$, $p_2^2 = m^2$, the electromagnetic current is j_μ and the neutron field is ψ, with $D_x = i(\gamma_\mu \overleftarrow{\partial}_\mu + m)$ and $\bar{\chi}(x) = \bar{\psi}(x) D_x$.

The neutron electric dipole moment is then given by

$$e\beta = \frac{1}{\pi} \int\limits_{(m+\mu)}^{\infty} \left\{ \frac{\operatorname{Re} A(W)}{W - m} + \frac{\operatorname{Re} A(-W)}{W + m} \right\} dW \qquad (5.3)$$

where $A(W)$ is found from $A_\mu(p_2, p_1)$ with $(p_1 - p_2)^2 = 0$ by the use of a projection operator

$$A(W) = \sum_{\text{Spin}} A_\mu(p_2, p_1) \frac{m}{2(W^2 - m^2)^2} (\gamma \cdot p_1 + W) \sigma_{\mu\nu}(p_1 - p_2)_\nu \gamma_5 n(p_2). \quad (5.4)$$

The next step is to insert a complete set of physical states into the matrix element of Eq. (5.2) and assume that meson-baryon intermediate states are the dominant contribution to the dispersion integral of Eq. (5.3). In this approximation the absorptive part is given by

$$A_\mu(p_2, p_1) = \frac{1}{2(2\pi)^2} \left(\frac{p_{20}}{m}\right)^{\frac{1}{2}} \int d^3 p_3 d^3 q \sum_{\text{Spin}} \delta(p_3 + q - p_1) \qquad (5.5)$$
$$\cdot \langle p_2 | j_\mu(0) | p_3 q \rangle \langle p_3 q | \bar{\chi}(0) | 0 \rangle$$

where p_3 and q are the momenta of the baryon and meson respectively. The two matrix elements of Eq. (5.5) correspond to meson photoproduction and the coupling of an off-shell neutron to a baryon and a meson. We shall take these latter to be a nucleon and a pion, and the photoproduction part to be dominated by the $P_{33}(1236)$ resonance where again we can insert the T-violating phase ϕ.

Parity violation is introduced by the weak interaction at the $np\pi^-$ vertex, the weak coupling constant involved being determined, up to a sign, by the S-wave non-leptonic decay rates. It turns out to be very small since it depends on the Cabbibo angle θ_c through a factor $\sin^2 \theta_c$.

Combining the two effects, Broadhurst obtained a value of

$$|\beta| = 4 \cdot 10^{-23} \sin\phi \text{ cm} . \tag{5.6}$$

Taking the typical T-violating phases found by Donnachie and Shaw [31] in their analysis of pion photoproduction this leads to a value of

$$|\beta| \simeq 5 \cdot 10^{-24} \text{ cm}$$

compared with an experimental upper limit of $5 \cdot 10^{-23}$ cm [56].

6. Conclusion

Existing pion photoproduction and π^- radiative capture data can readily be interpreted in terms of C-violating and isotensor components in the electromagnetic current. The C-violation comes from a direct experimental test and theoretical argument is involved only as a means of quantifying the effect in terms of a phase. The argument for an isotensor current is more involved and different theoretical prejudices lead to different conclusions about its existence.

The magnitude of the C-violation observed does not lead to any disagreement with other experimental data. In particular it is consistent with deuteron photodisintegration and neutron proton radiative capture, inelastic electron scattering off a polarized target and the neutron electric dipole moment. However these tests are rather insensitive to C-violation, since in each case they depend on the interference of the C-violating part with a small amplitude, leading to a considerable suppression of the effect. Additionally none of these reactions can say anything about an isotensor electromagnetic current.

Thus we are left with pion photoproduction, and its inverse, as the only suitable reactions. To establish C-violation more π^- radiative capture data is required and a clarification of the photoproduction data off deuterium. The most promising experiment here is to measure π^-/π^+ ratios with a tagged photon beam. For an isotensor current, the only unambiguous test is to compare $\gamma n \to \pi^0 n$ with $\gamma p \to \pi^0 p$ since the non-resonant background in each of these reactions is extremely small and they are completely dominated by the resonance. If there is no isotensor current, then the cross-sections should be the same. If there is an isotensor current of the magnitude obtained by Donnachie and Shaw [31] then the neutron cross section should be only about half the proton cross section.

References

1. Bernstein, J., Feinberg, G., Lee, T. D.: Phys. Rev. **139**, 1650 (1965).
2. Grishin, V. G., Lyuboshitz, V. L., Ogievetskii, V. I., Podgoretskii, M. I.: Yod. Fiz. (USSR) **4**, 126 (1966); Sov. J. Nucl. Phys. **4**, 90 (1967).
3. Dombey, N., Kabir, P. K.: Phys. Rev. Letters **17**, 730 (1966).
4. Shaw, G.: Nucl. Phys. B**3**, 338 (1967).
5. Sanda, A. I., Shaw, G.: Phys. Rev. Letters **24**, 1310 (1970); Phys. Rev. D**3**, 243 (1971).
6. Lee, T. D.: Phys. Rev. **140**, B 959 (1965).
7. — Phys. Rev. **140**, B 967 (1965).
8. Adler, S. L.: Phys. Rev. Letters **18**, 519 (1967).
9. Gormley, M., Hyman, E., Lee, W., Nash, T., Peoples, J., Schultz, C., Stein, S.: Phys. Rev. Letters **21**, 402 (1968).
10. Bultran, M. T., Kreisler, M. N., Mischke, R. E.: Phys. Rev. Letters **25**, 1358 (1970).
11. Gormley, M., Hyman, E., Lee, W., Nash, T., Peoples, J., Schultz, C., Stein, S.: Phys. Rev. Letters **21**, 399 (1968).
12. Thresher, J. G.: Proc. Daresbury Study Weekend, Jan. 1971 DNPL/R 9.
13. Herzo, D., Banner, D., Beier, E., Bertram, W. K., Edwards, R. T., Frauenfelder, H., Koester, L. J., Rosenberg, E., Russell, A., Segler, S., Wattenberg, A.: Phys. Rev. **186**, 1403 (1969).
14. Blin-Stoyle, R. J.: Phys. Rev. Letters **23**, 535 (1970).
15. Wilkinson, D. H.: Phys. Letters **12**, 348 (1964).
16. Snover, K. A., Heikkinen, D. W., Riess, F., Kuan, H. M., Hanna, S. S.: Phys. Rev. Letters **22**, 239 (1969).
17. Aachen-Berlin-Bonn-Hamburg-Heidelberg-München (ABBHHM) Collaboration, Nuclear Physics B**8**, 535 (1968), Butenschon, H., DESY RI-70/1 (1970).
18. Pavia-Frascati-Rome-Napoli (PFRN) Collaboration, Nuovo Cimento Letters **3**, 697 (1970).
19. Garelick, D., Cooperstein, G.: Phys. Rev. **136**, B 201 (1964).
20. Fujii, T., Okuno, H., Orito, S., Sasaki, H., Nozaki, T., Takasaki, F., Takikawa, K., Amako, K., Endo, I., Yoshida, K., Higuchi, M., Sato, M., Sumi, Y.: Phys. Rev. Letters **26**, 1672 (1971).
21. Hoog, W. R.: Proc. Phys. Soc. (Lond.) **80**, 729 (1962).
22. Favier, J., Alder, J. C., Joseph, C., Vaucher, B., Schinzel, D., Zupancic, C., Bressani, T., Chiavassa, E.: Phys. Letters **31** B, 609 (1970).
23. Schinzel, D.: Private communication.
24. Berardo, P. A., Haddock, R. P., Nefkens, B. M. K., Verhey, L. J., Zeller, M. E.: Phys. Rev. Letters **24**, 419 (1970); **26**, 201 (1971).
25. — — — — — Phys. Rev. Letters **26**, 205 (1971).
26. White, D. H., Schectman, R. M., Chasan, B. M.: Phys. Rev. **120**, 614 (1960).
27. Noelle, P., Pfeil, W., Schwela, D.: Bonn Preprint P I 2–79 (1970).
28. Berends, F. A., Wearer, D. L.: CEA Preprint (1970).
29. Christ, N., Lee, T. D.: Phys. Rev. **148**, 1520 (1966).
30. Sanda, A. I., Shaw, G.: Phys. Rev. Letters **26**, 1057 (1971). Columbia University Preprint NYO-1932(2)-197(1971).
31. Donnachie, A., Shaw, G.: Phys. Letters (to be published) Daresbury Preprint DNPL (P 79).
32. Chew, G. F., Goldberger, M. L., Low, F. E., Nambu, Y.: Phys. Rev. **106**, 1345 (1957).
33. Watson, K. M.: Phys. Rev. **85**, 852 (1952).
34. Donnachie, A.: Proc. Scottish Universities Summer School, p. 109. New York: Academic Press 1971.
35. Berends, F. A., Donnachie, A., Weaver, D. L.: Nucl. Phys. B**4**, 54 (1968).
36. Devenish, R. C. E., Lyth, D. H., Rankin, W. A.: University of Lancaster Preprint, 1971.

37. — — — Daresbury Preprint (in preparation).
38. Keck, J. C., Tollestrup, A. V.: Phys. Rev. **101**, 360 (1956).
39. Whalin, E. A., Schriever, D., Hanson, A. O.: Phys. Rev. **101**, 377 (1956).
40. Barshay, S.: Phys. Rev. Letters **17**, 49 (1966).
41. Austern, A.: Phys. Rev. **100**, 1522 (1955).
42. Sober, D. I., Cassel, D. G., Sadoff, A. J., Chen, K. W., Crean, P. A.: Phys. Rev. Letters **22**, 430 (1969).
43. Buon, J., Gracco, V., Lefrancois, J., Lehmann, P., Merkel, B., Roy, Ph.: Phys. Letters, **26 B**, 595 (1968).
44. Anderson, R. L., Prepost, R., Wiik, B. H.: Phys. Rev. Letters **22**, 651 (1969).
45. Shrock, B. L., Haddock, R. P., Helland, J. A., Longo, M. J., Wilson, S. S., Young, K. K., Cheng, D., Perez-Mendez, V.: UCLA 34 P 106-53.
46. Christ, N., Lee, T. D.: Phys. Rev. **143**, 1310 (1966).
47. Albrecht, W., *et al.*: Nucl. Phys. B **25**, 1 (1971).
48. Hellings, R. D., *et al.*: Daresbury Preprint DNPL/P 65 (1971).
49. Gayler, G.: Proceedings Daresbury Study Weekend on Inelastic Electron Scattering.
50. Chen, J. R., *et al.*: Phys. Rev. Letters **21**, 1279 (1968).
51. Rock, S., *et al.*: Phys. Rev. Letters **24**, 748 (1970).
52. Bincer, A. M.: Phys. Rev. **118**, 855 (1960).
53. Drell, S. D., Pogels, H. R.: Phys. Rev. **140**, B 397 (1965).
54. Barton, G., White, E. D.: Phys. Rev. **184**, 1660 (1969).
55. Broadhurst, D. J.: Nucl. Phys. B **20**, 603 (1970).
56. Baird, J. K., Miller, P. D., Dress, W. B., Ramsey, N. F.: Phys. Rev. **179**, 1285 (1969).

Professor Dr. A. Donnachie
Department of Theoretical Physics
University of Manchester
Manchester, M 13 9 PL
England

Regge Analysis and Dual Absorptive Model

A. P. CONTOGOURIS

Contents

1. Introduction

The momentum-transfer structure of two-body reactions has often given rise to important tests of high energy models. This is particularly true for the Regge model that owes many of its successes (and failures) to its ability (or inability) to offer simple explanations of this structure.

In its original version the Regge model attributes the dips observed in twobody reactions to nonsense wrong signature zeros associated with the Regge pole exchange; it also admits Regge cuts but, in general, of small magnitude (weak cut model) [1]. During the last two years an alternative Regge model has been advanced; this explains the dips via destructive interference between the Regge pole and a strong Regge cut, whose presence is related to absorptive effects (strong cut reggeized absorption model) [2]. It is now recognized that each of these models is only partly successful, facing serious difficulties in several reactions.

In an effort to combine the successes of each of these approaches Harari has recently proposed the dual absorptive model [3, 4]. Although this should be considered as a rather qualitative tool, the subsequent discussion will show that, within the Regge approach, it can also be used as a guideline for quantitative estimates and detailed phenomenological analysis.

Section 2 reviews briefly the weak cut model together with some of its most serious difficulties. Section 3 reviews the strong cut reggeized absorption model and some of its failures. Section 4 presents the dual absorptive model and its resolution of a number of difficulties. Finally, Section 5 discusses a Regge pole-Regge cut analysis of certain important quasi-elastic reactions and shows how the dual absorptive model can be used in detailed calculations.

2. The Weak Cut Model

We consider the nondiffractive reaction meson(a) + nucleon(b)→meson(c) + baryon(d) (Fig. 1) and denote the s-channel helicity amplitude (SHA) by

$$M_{\lambda_c - \lambda_d, \lambda_a - \lambda_b} \equiv M_{\lambda, \mu}(s, t) ; \qquad (2.1)$$

in cases of elastic scattering we shall be interested in the t-structure of the nondiffractive part.

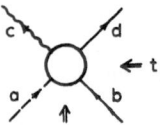

Fig. 1. The reaction meson(a) + nucleon(b) → meson(c) + baryon(d)

The WCM starts with Regge residues involving nonsense-wrong-signature zeros (NWSZ). In fact, a consise form involves simple Veneziano residues: the exchange of the Regge pole R with trajectory $\alpha_R = \alpha_R(t)$ is:

$$M_{\lambda, \mu}^{(R)}(s, t) = (-t)^{(n+x)/2} \frac{G}{\Gamma(\alpha_R)} \frac{1 \pm \exp - i\pi\alpha_R}{\sin \pi\alpha_R} (\alpha's)^{\alpha_R} . \qquad (2.2)$$

Here

$$n = |\lambda - \mu| = |\lambda_c - \lambda_d - (\lambda_a - \lambda_b)| \qquad (2.3)$$

is the overall helicity flip; the factor $(-t)^{n/2}$ expresses angular momentum conservation. Also

$$x = |\lambda_c - \lambda_a| + |\lambda_d - \lambda_b| - n \qquad (2.4)$$

and the factor $(-t)^{x/2}$ expresses parity conservation (or, equivalently, the assumption that the exchanged trajectory is nonconspiring) [3]. G is a constant given by:

$$G = g_{ca} g_{bd} \qquad (2.5)$$

where g_{ca} is the coupling of the lowest physical particle of the trajectory $\alpha_R(t)$ to the external particles c, a. Finally, α' is the slope of the trajectory.

In addition to the Regge pole (2.2) there is also a Regge cut $M_{\lambda, \mu}^{(RP)}$ generated by inserting (2.2) in a Bessel-Hankel transform of the type [1, 2]:

$$M_{\lambda, \mu}^{(RP)}(s, t) \sim \int\limits_0^\infty d\tau J_n(\tau^{\frac{1}{2}}(-t)^{\frac{1}{2}}) M_{\lambda, \mu}^{(R)}(s, \tau) M^{(P)}(s, \tau) . \qquad (2.6)$$

Here $M^{(P)}(s, \tau)$ represents the Pomeron exchange or elastic scattering properly parametrized; thus $M^{(RP)}_{\lambda, \mu}$ represents a Reggion-Pomeron cut. Notice that for fixed s $M^{(R)}_{\lambda, \mu}$ and $M^{(P)}$ decrease exponentially with τ; then the above integral receives most of its contribution from the interval $0 < \tau \lesssim 1$ GeV^2. As said, in the WCM $M^{(R)}_{\lambda, \mu}$ contains NWSZ. For a vector exchange (ϱ, ω) this implies that $M^{(R)}_{\lambda, \mu}(s, \tau)$ changes sign at $\tau \approx 0.55$ GeV^2. Hence the integral (2.6) contains two important pieces with opposite sign and the resulting cut contribution is relatively weak (WCM).

The WCM, although quite successful, encounters certain difficulties, the most important of which are:

(2.A). Crossover [5]: Consider the differential cross-sections for $K^{\mp} p \to K^{\mp} p$. At all high energies they cross each other $at - t = -t_c \simeq 0.15 \sim 0.20$ GeV^2. The most important contributions to these processes are:

$$T_{K^{\mp} p} = P + f \pm \omega \qquad (2.7)$$

where e.g. ω means the t-channel exchange with the ω quantum numbers. From (2.7) the difference between $K^- p$ and $K^+ p$ is due to ω; hence the overall ω exchange must change sign at $t = t_c$. In models with Regge *poles* only, a zero at $t = t_c$ was arbitrarily introduced in the residue function of ω. However, isospin invariance implies that the physical quantity

$$X_\omega(s, t) = \frac{d\sigma}{dt} (\pi^+ p \to \varrho^+ p) + \frac{d\sigma}{dt} (\pi^- p \to \varrho^- p) - \frac{d\sigma}{dt} (\pi^- p \to \varrho^0 n) \quad (2.8)$$

contains only exchange with the ω quantum numbers [6]. If ω exchange in $K^\pm p \to K^\pm p$ has a zero at $t = t_c$, then factorization implies that also in $\pi N \to \varrho N$ the ω exchange must have a zero. Analysis of available data at 4 and 8 GeV gives no such evidence at all (Fig. 11). In contrast, it gives evidence for a zero at $t = -0.55$ GeV^2. – We thus need an additional contribution with the ω quantum numbers. The best candidate is an ωP Regge cut. However, this cut must have the following properties:

(i) In $\pi N \to \varrho N$ it must be relatively weak so that the zero at $t = -0.55$ of the ω Regge pole remains essentially unshifted

(ii) In $K^\pm p \to K^\pm p$ it must be very strong so that the zero is shifted from $t = -0.55$ to $t = t_c$. This cannot be done in the WCM.

(2.B) Absence of Dips in $\pi N \to \omega N$, $\pi N \to \omega \Delta$, $\gamma p \to \eta^0 p$ and $\frac{d\sigma}{dt} (\gamma p \to \pi^+ n) - \frac{d\sigma}{dt} (\gamma n \to \pi^- p)$. In brief, all these reactions are dominated by ϱ exchange and the WCM predicts a dip at $t = -0.55$. However, experiment in no case shows a dip.

3. The Strong Cut Reggeized Absorption Model

In the SCRAM the Regge pole residues do not contain NWSZ. The same form (2.6) is used to generate the Regge cut. However, now the pole contribution has the same sign throughout $0 < \tau \lesssim 1 \text{ GeV}^2$ and the resulting cut is significantly stronger than before. Furthermore it is multiplied by an arbitrary factor $\tilde{\lambda}$ ("coherent inelastic") [2] that ranges from 1 to 3.5. Thus e.g. the dip in $\pi^- p \to \pi^0 n$ is produced through destructive interference between the ϱ Regge pole and a very strong ϱP cut.

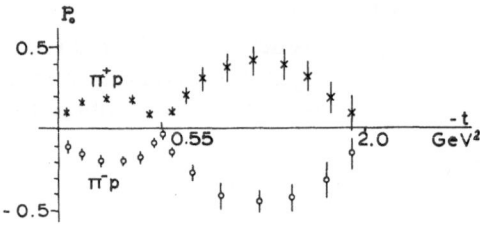

Fig. 2. The t-dependence of the data for polarization P_0 in $\pi^\pm p \to \pi^\pm p$

SCRAM is successful in describing the crossover as well as the reactions (2.B). However, it fails in the following:

(3.A). Polarization in $\pi^\pm p \to \pi^\pm p$. The basic features of the data for polarization P_0 in $\pi^+ p$ and $\pi^- p$ elastic scattering can be summarized as follows (Fig. 2).

(i) Near mirror symmetry.

(ii) At $t \simeq -0.55: P_0(\pi^\pm p) = 0$.

(iii) $P_0(\pi^\pm p) \gtrless 0$ everywhere.

Similar features (although not so clear-cut) are present in the polarization data for $K^\pm p \to K^\pm p$. — Let f be the spin nonflip and g the spin flip part of the πN amplitude, so that

$$P_0 \sim \text{Im}(f^* g). \tag{3.1}$$

The simplest description assumes that f is dominated by Pomeron exchange (so that for small $|t|$ $f \approx$ imaginary) and that g is dominated by exchange of ϱ (or $\varrho + \varrho P$ cut), so that

$$P_0 \sim f_P \cdot \text{Reg}_\varrho. \tag{3.2}$$

Since ϱ exchange contributes with opposite sign in $\pi^+ p$ and $\pi^- p$, the feature (i) is easily understood in both the WCM and the SCRAM. Since $\text{Reg}_\varrho \approx 0$ at $t = -0.55$ in both WCM and SCRAM, feature (ii) is also easily understood. However, SCRAM fails completely with respect to feature (iii). As said, in this model the dip of $\pi^- p \to \pi^0 n$ at $t = -0.55$ is explained through destructive interference between the ϱ pole and a ϱP cut. Hence

$$g(s, t) = g_\varrho(s, t) - g_{\varrho P}(s, t) \tag{3.3}$$

so that

$$\text{Reg}(s, t) = \text{Reg}_\varrho(s, t) - \text{Reg}_{\varrho P}(s, t) \gtrless 0 \quad \text{for} \quad -t \gtrless 0.55 \tag{3.4}$$

thus, in varying through $t = -0.55$, P_0 should change sign for each of $\pi^+ p$ and $\pi^- p$. This is in clear contradiction with experiment. – Notice that in contrast the WCM gives

$$\text{Reg}(s, t) \simeq \text{Re}\left[b\alpha_\varrho(1 - e^{-i\pi\alpha_\varrho}) s^{\alpha_\varrho}/\sin \pi \alpha_\varrho\right]$$
$$= b\alpha_\varrho(1 - \cos \pi \alpha_\varrho) s^{\alpha_\varrho}/\sin \pi \alpha_\varrho \xrightarrow[\alpha_\varrho \to 0]{} \tfrac{1}{2} b\pi \alpha_\varrho^2 s^{\alpha_\varrho}, \tag{3.5}$$

i.e. Reg (and P_0) has the same sign before and after $\alpha_\varrho = 0$ (i.e. $t = -0.55$).

(3.B). Absence of Dips in $\pi^- p \to \eta^0 n$, $\pi^+ p \to \eta^0 \Delta^{++}$ and $K^- p \to \bar{K}^0 n$, $K^\perp n \to K^0 p$. These reactions are dominated by A_2 or $\varrho + A_2$ exchange. There is much evidence that in all of them the helicity flip ($n = 1$) amplitude dominates. The evidence comes from: (i) the dipping or rounding of $d\sigma/dt$ near $t = 0$, (ii) FESR for the KN change exchange reactions, and (iii) exchange degeneracy between ϱ and A_2 plus the fact that $\pi^- p \to \pi^0 n$ is dominated by helicity flip. In SCRAM the ϱ and A_2 residues have the same structure. With dominant helicity flip SCRAM then predicts $A_2 P$ and ϱP cuts of the same magnitude; hence for all above reactions the same structure as $\pi^- p \to \pi^0 n$. This also contradicts experiment.

4. The Dual Absorptive Model (DAM) [4]

Suppose that $M_{\lambda, \mu}(s, t)$ is the *complete* SHA for the reaction $ab \to cd$ (Fig. 1). We assumed that the t-channel of this reaction does not admit Pomeron exchange (nondiffractive). Consider the impact parameter representation:

$$M_{\lambda, \mu}(s, t) \sim s \int_0^\infty b\,db\,\mathscr{M}_{\lambda, \mu}(s, b)\,J_n(b\sqrt{-t}). \tag{4.1}$$

The DAM is defined by the Ansatz:

$$\text{Im}\,\mathscr{M}_{\lambda, \mu}(s, b) = c_{\lambda, \mu}\delta(b - R) \quad R \simeq 1 \text{ fermi}; \tag{4.2}$$

then

$$\text{Im}\,M_{\lambda, \mu}(s, t) = C_{\lambda, \mu}(s)\,J_n(R\sqrt{-t}) \tag{4.3}$$

where $C_{\lambda, \mu}(s)$ is a known function of s.

In view of the fact that $ab \to cd$ is nondiffractive, Eq. (4.2) can be seen to result from the assumption that at energy s $\operatorname{Im} M_{\lambda,\mu}(s,t)$ is locally dominated by resonances of mass $\approx \sqrt{s}$; this is a statement of duality. At the energies of interest ($5 \lesssim E_{\mathrm{lab}} \lesssim 20$ GeV) these are resonances of high spin. Assuming, as usual, a hadron interaction radius $R \simeq 1$ fermi, their spin is roughly

$$J + \tfrac{1}{2} \approx R q_s \quad \text{(impact parameter } b = (J + \tfrac{1}{2})/q_s); \tag{4.4}$$

thus these resonances populate the "peripheral" partial waves; and this essentially implies Eq. (4.2). Notice that for $\operatorname{Im} M_{\lambda,\mu}(s,t)$ this conclusion is equivalent to the well-known statement of absorption. However, in contrast to absorption, duality makes no simple statement about $\operatorname{Re} M_{\lambda,\mu}(s,t)$; thus in the DAM $\operatorname{Re} M_{\lambda,\mu}$ does not, in general, have the form $J_n(R \sqrt{-t})$.

In the determination of $\operatorname{Re} M_{\lambda,\mu}(s,t)$ it is important to keep track of an application of the Phragmén-Lindelöff (PL) theorem concerning the phase of the terms in asymptotic expansions of crossing symmetric amplitudes [7]: Suppose that

$$F(s,t) = \sum_i \beta_i(t)\, s^{\alpha_i(t)}/(\log \log \dots s)^{\gamma_i(t)} \tag{4.5}$$

log log ... is an arbitrary but finite number of log) is such an expansion. Then the PL theorem implies that the phase of the leading contribution (to log log ... s) in the term i is determined by the factor $1 \pm \exp - i\pi \alpha_i(t)$ where $\alpha_i(t)$ is precisely the exponent of s (the $\pm$ sign according to whether the amplitude is crossing-even or odd).

Now, in general $M_{\lambda,\mu}$ will contain a pole plus a cut:

$$M_{\lambda,\mu}(s,t) = M_{\lambda,\mu} + M_{\lambda,\mu}^{(RP)};$$

the pole will have its signature factor $1 \pm \exp - i\pi \alpha_R$; the cut will also have its signature factor $1 \pm \exp - i\pi \alpha_{RP}$, where

$$\alpha_{RP}(t) = \alpha_R(0) + \frac{\lambda_R \lambda_P}{\lambda_R + \lambda_P}\, t .$$

Then the real part will be determined uniquely and the PL theorem will be satisfied automatically.

To make clear how the DAM accounts for the experimental situation and also how it gives an estimate of the magnitude of the Regge cuts we shall discuss a number of special cases. We start with the helicity flip amplitude ($n = 1$), when, as will become clear, DAM $\equiv$ WCM. For $n = 1$ the DAM requires:

$$\operatorname{Im} M_{\lambda,\mu}(s,t) \sim J_1(R \sqrt{-t}) \tag{4.6}$$

leading to the indicated t-structure: one zero at $t = 0$ and another at $t \simeq -0.55 \, \mathrm{GeV}^2$ (for $R \simeq 1$ fermi) (Fig. 3).

(4.A) ϱ-Exchange in $\pi^- p \to \pi^0 n$. Here helicity flip dominates and the indices n, x of (2.2) take the values $n = 1$ $x = 0$. We proceed with the assumption that the Regge pole exchanges contain NWSZs. Then the imaginary part of the Regge pole has the form

$$\operatorname{Im} M_{\lambda,\mu}^{(\varrho)}(s, t) = (-t)^{(n+x)/2} \frac{G}{\Gamma(\alpha_\varrho)} \frac{(\alpha' s)^{\alpha_\varrho(t)}}{\sin \pi \alpha_\varrho} \operatorname{Im}(1 - \exp - i\pi \alpha^\varrho)$$

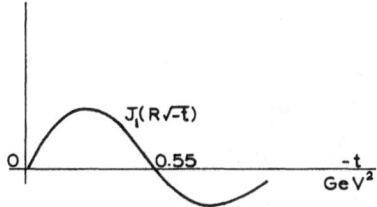

Fig. 3. Form of the Bessel-function $J_1(R\sqrt{-t})$ for $R \simeq 1$ fermi

and for $-t \lesssim 1 \, \mathrm{GeV}^2$

$$\simeq (-t)^{\frac{1}{2}} G \alpha_\varrho(t) (\alpha' s)^{\alpha_\varrho(t)}. \tag{4.7}$$

This already has the form of $J_1(R\sqrt{-t})$. Conclusion: to this ϱ Regge pole exchange add weak (or no) cut. – Thus, for the complete $M_{\lambda,\mu}(s, t)$ we have essentially to multiply (4.7) with the signature factor $1 - \exp - i\pi \alpha_\varrho$. Clearly, this will lead to correct $\pi^\pm p$ polarizations (see end of (3.A)).

(4.B) ω-Exchange in $\pi N \to \varrho N$ (related through Vector Dominance to $\gamma p \to \pi^0 p$). The ωNN coupling is mainly nonflip; thus the $n = 1$ amplitude dominates and the considerations of (4.A) apply as well. An ω Regge pole containing NWSZ results in a t-structure of the form $J_1(R\sqrt{-t})$. Again, this implies a weak cut. Clearly, $\gamma p \to \pi^0 p$ and $X_\omega(s, t)$ must have a dip at $t = -0.55$; this is indeed observed (see Section 5 and Figs. 7 and 11).

(4.C) A_2-Eschange in $\pi^- p \to \eta^0 n$. Again this is dominated by the $n = 1$ amplitude and the DAM demands the t-structure of (4.6). An A_2 Regge pole with NWSZ will again give

$$\operatorname{Im} M_{\lambda,\mu}^{(A_2)}(s, t) \approx (-t)^{\frac{1}{2}} G \alpha_{A_2}(t) (\alpha' s)^{\alpha_{A_2}(t)} \tag{4.8}$$

thus satisfying (4.6). The conclusion is again: weak (or no) cuts. To obtain the complete amplitude we must multiply (4.8) by $1 \pm \exp - i\pi \alpha_{A_2}$. Then

$$\operatorname{Re} M_{\lambda,\mu}^{(A_2)}(s, t) = (-t)^{\frac{1}{2}} G \alpha_{A_2}(t) \frac{(\alpha' s)^{\alpha_{A_2}}}{\sin \pi \alpha_{A_2}} \operatorname{Re}(1 + \exp - i\pi \alpha_{A_2})$$

$$\xrightarrow[\alpha_{A_2} \to 0]{} (-t)^{\frac{1}{2}} G(\alpha' s)^{\alpha_A} (2/\pi) \neq 0. \tag{4.9}$$

Thus the DAM predicts no dip at $\alpha_A(t) = 0$ in accord with experiment. — An important point to notice is that the $\mathrm{Re}\, M_{\lambda,\mu}$ of (4.9) has no zero in $0 \leqq -t < 1 \ \mathrm{GeV}^2$; thus it does not bear any resemblance with $J_1(R\sqrt{-t})$. Clearly, $\mathrm{Re}\, M_{\lambda,\mu}(s,t)$ ist not dominated by the peripheral partial waves, at least in this case.

We turn now to the helicity nonflip amplitude ($n = 0$) when the DAM requires

$$\mathrm{Im}\, M_{\lambda,\mu}(s,t) \sim J_0(R\sqrt{-t}) ; \tag{4.10}$$

with $R \simeq 1$ fermi this leads to the t-structure of Fig. 4.

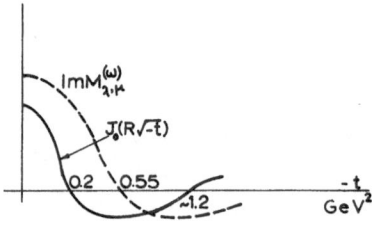

Fig. 4. Form of the Bessel function $J_0(R\sqrt{-t})$ for $R \simeq 1$ fermi and of $\mathrm{Im}\, M_{\lambda,\mu}^{(\omega)}$ of Eq. (4.11)

(4.D) ω-Exchange in $K^{\pm}p \to K^{\pm}p$ (crossover). With the ωNN coupling mainly nonflip this exchange is dominated by the amplitude with $n = 0$. Here, the indices of (2.2) are $n = x = 0$. Starting with an ω Regge pole containing NWSZ we have

$$\mathrm{Im}\, M_{\lambda,\mu}^{(\omega)}(s,t) = \frac{G}{\Gamma(\alpha_\omega)} \frac{(\alpha' s)^{\alpha_\omega(t)}}{\sin \pi \alpha_\omega} \mathrm{Im}(1 - \exp - i\pi\alpha_\omega) \approx G\alpha_\omega(t)(\alpha' s)^{\alpha_\omega(t)}. \tag{4.11}$$

This has a zero at $t = -0.55$ and thus a structure different from $J_0(R\sqrt{-t})$. DAM implies that a strong ωP cut must be added; its contribution must be of sign opposite to the ω-pole, so that the zero is shifted to $t = -0.2$. A cut calculated by inserting (4.11) in Bessel-Hankel transforms of the type (2.6) is not sufficiently strong; its magnitude must be multiplied by a factor $\tilde{\lambda} \approx 2$. This is an "absorptive" cut in all respects similar to these of SCRAM; and accounts well for all the experimental crossovers ($K^{\pm}p$, $\bar{p}p$ with pp, $\pi^{\pm}p$). Now, the complete nonflip amplitude will be

$$M_{\lambda,\mu}(s,t) = G\alpha_\omega(t) \frac{(\alpha' s)^{\alpha_\omega(t)}}{\sin \pi \alpha_\omega}(1 - e^{-i\pi\alpha_\omega}) - C_{\omega P} \frac{(\alpha' s)^{\alpha_{\omega P}(t)}}{\log \alpha' s}(1 - e^{-i\pi\alpha_{\omega P}}).$$

Clearly, the pole and cut signature factors uniquely determine the real part of the amplitude, as well. Moreover, the phase of each term is in accord with the PL theorem.

5. Regge Analysis and DAM for Certain Quasi-Elastic Reactions

This section will take up the reactions

$$\pi^+ n \to \omega^0 p, \quad \pi^+ p \to \omega^0 \Delta^{++} \qquad (5.1\text{--}2)$$

$$\gamma p \to \eta^0 p \qquad (5.3)$$

and the cross-section difference

$$\frac{\mathrm{d}\sigma}{\mathrm{d}t}(\gamma p \to \pi^+ n) - \frac{\mathrm{d}\sigma}{\mathrm{d}t}(\gamma n \to \pi^- p); \qquad (5.4)$$

we mentioned in (2.B) that these present special difficulties in the WCM. The ϱ exchange that dominates these reactions is also important in

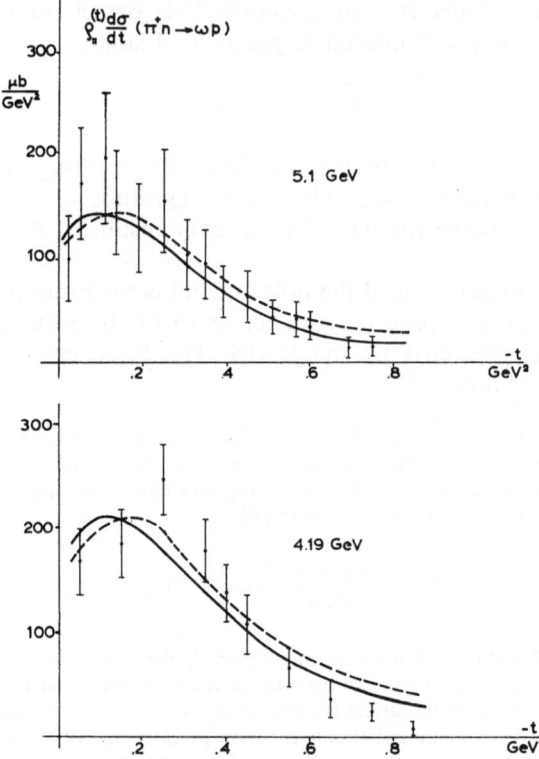

Fig. 5. The quantity $\varrho_{11}^{(t)}\mathrm{d}\sigma/\mathrm{d}t$ for $\pi^+ n \to \omega p$ ($\varrho_{mn}^{(t)}$ = density matrix elements in Gottfried-Jackson frame). – In all Figs. 5–12: Broken lines: model A (see Footnote on p. 154); continuous lines: model B. Data B-B-F-O Collaboration [15]

understanding the details of the photo-reactions

$$\gamma p \to \pi^0 p \qquad \gamma n \to \pi^0 n \qquad\qquad (5.5\text{--}6)$$

However, in (5.5) an (5.6) ω exchange plays a dominant role; and once this is determined, vector dominance predicts immediately the magnitude of $X_\omega(s, t)$ of (2.8). Thus a meaningful Regge pole – Regge cut analysis should simultaneously encompass all reactions (5.1)–(5.6) and (2.8). Here, we shall report on a recent analysis [8] that also makes comparison with DAM.

This analysis starts with Regge pole residues involving NWSZs and in fact of the Veneziano form (2.2). The residue constants (2.5) are determined, at least approximately (within factors of 2), in terms of known couplings and from the combined requirements of exchange degeneracy, vector meson dominance and $SU(6)_W{}^1$; these essentially fix the magnitude of the Regge pole contributions.

In addition, Regge cuts are introduced via Bessel-Hankel transforms (2.6) [2] and using a Pomeron Regge pole of slope

$$\alpha_P' = 0.3 \text{ GeV}^{-2}. \qquad\qquad (5.7)$$

In certain cases (to be discussed below) the resulting contribution is multiplied by a factor $\tilde\lambda \approx 2$. The exact magnitude of $\tilde\lambda$ in each case is determined by fitting the data. Preliminary results of the fits are given in Figs. 5–11 [2].

Having thus determined the pole and cut contributions, at each fixed energy one can compare the t-dependence of the resulting imaginary parts to that predicted by the DAM. The basic conclusions can be summarized as follows:

---

[1] $SU(6)_W$ is important to predict $\pi^+ p \to \omega^0 \varDelta^{++}$ in terms of $\pi^+ n \to \omega^0 p$ as well as the ϱ and ω exchanges in $\gamma p \to \eta^0 p$ in terms of ϱ and ω in $\gamma N \to \pi^0 N$.

[2] Due to the presence of NWSZs in the Regge residues, the complete ϱP cut (or ωP cut) calculated from Ref. [2] is of the form [8]

$$A_{\varrho P}(s, t) = \frac{1 - \exp - i\pi\alpha_{\varrho P}}{(\log \alpha' s)^k} (\alpha' s)^{\alpha_{\varrho P}(t)} (b_1 - b_2/\log \alpha' s) \qquad (A)$$

where $k = 1$ or 2 and b_1, b_2 known quantities (weakly dependent on t). Two models have been considered differing in the way the $\tilde\lambda$ factor is introduced: *Model A* (broken lines of Figs. 5–11): The $\tilde\lambda$ factor multiplies the overall $A_{\varrho P}$ of (A). *Model B* (continuous lines of Figs. 5–11): The $\tilde\lambda$ factor multiplies only the term proportional to b_2 in $A_{\varrho P}$. Notice that this term corresponds to a nonasymptotic contribution (the term b_1 dominates in the limit logs $\to \infty$); hence model B tends asymptotically to the WCM [9]. Notice that most of the conclusions about Regge cuts from Feynman graphs or dual loops hold to the leading order in logs. Trajectory slopes used in the fits: $\alpha_\varrho' = \alpha_\omega' = \alpha_{A_2}' = 0.9 \text{ GeV}^{-2}$, $\alpha_B' = 0.7 \text{ GeV}^{-2}$.

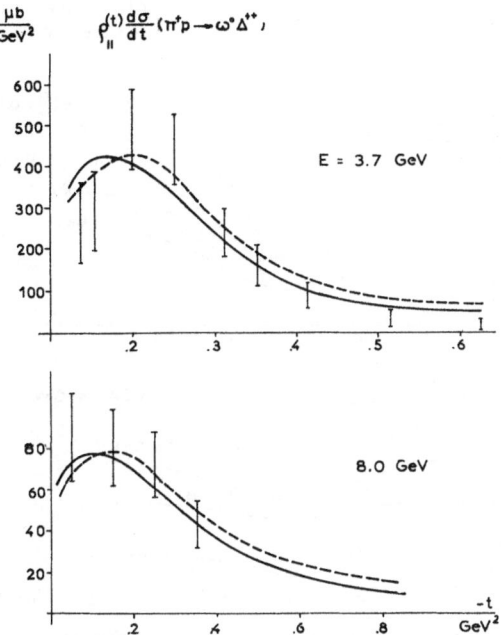

Fig. 6. The quantity $\varrho_{11}^{(t)} \, d\sigma / dt$ for $\pi^+ p \to \omega \Delta^{++}$. Notation as in Fig. 5. Data: Ref. [16]

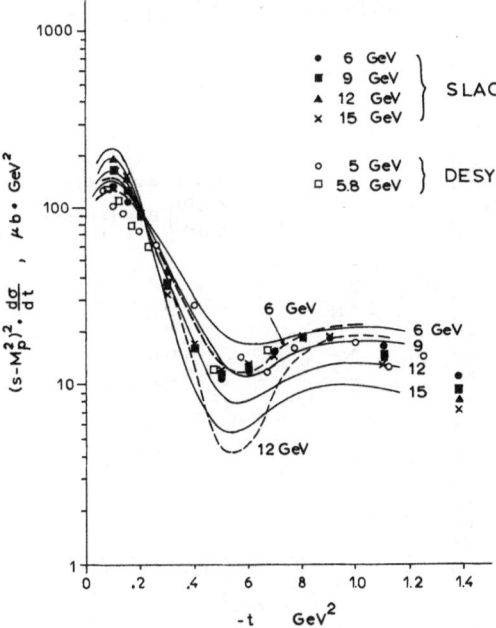

Fig. 7. Differential cross secions for $\gamma p \to \pi^0 p$

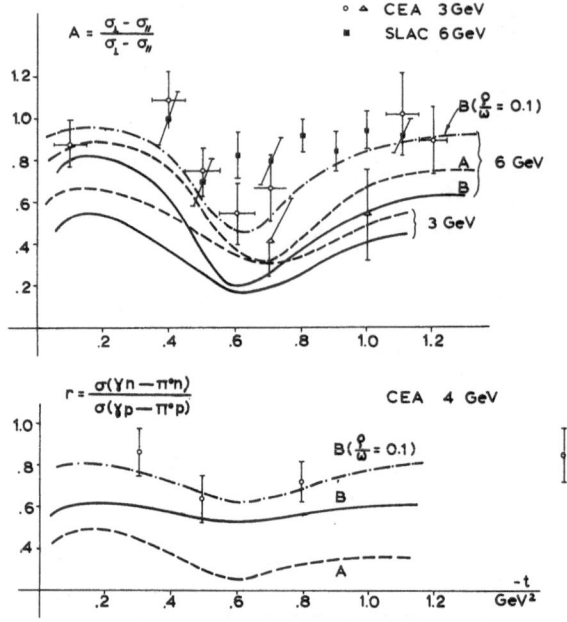

Fig. 8. Assymmetry ratio A for $\gamma p \to \pi^0 p$ and ratio r of differential cross-sections for π^0 photoproduction on neutrons and on photons. Broken lines: model A. Continuous and dash-dot lines: model B with different percentages of ϱ/ω exchange contributions (ϱ/ω = 0.15 and 0.1 respectively [8])

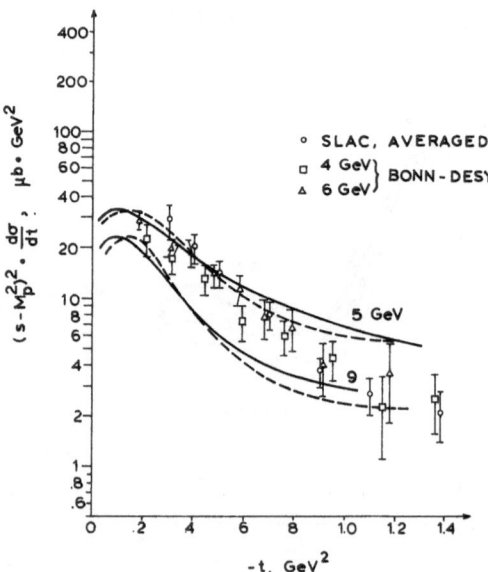

Fig. 9. Differential cross-sections for $\gamma p \to \eta^0 p$

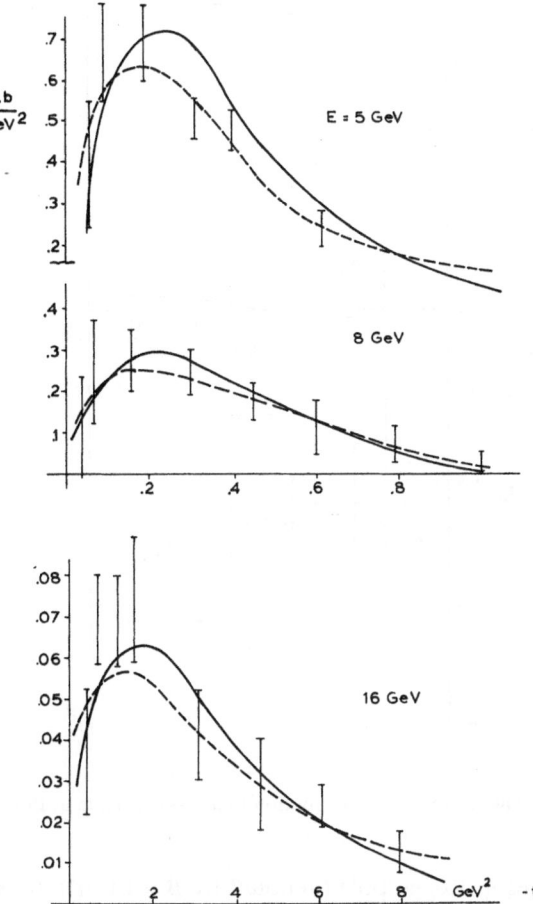

Fig. 10. The difference of differential cross-sections for $\gamma p \to \pi^+ n$ and $\gamma n \to \pi^- p$. Data: Ref. [17]

(5.A) Dominant Part of ω-Exchange: As said in (4.B), this is helicity flip and the DAM predicts a t-structure like $J_1(R\sqrt{-t})$. The imaginary parts of the amplitudes at two different energies (5.1 and 12 GeV) are presented in Fig. 12; they indeed exhibit the structure of $J_1(R\sqrt{-t})$. Notice that the cut contributions to these amplitudes have been multiplied by $\tilde{\lambda}$ factors of magnitude $\tilde{\lambda} \approx 1$ (WCM).

(5.B) Dominant Part of ϱ-Exchange: Since the ϱNN coupling is mainly flip, the $n=1$ amplitude of the ϱ exchange contribution to (5.1)–(5.6) is relatively small. We thus first consider the $n=0$ and 2 amplitudes and denote by A_ϱ and $A_{\varrho P}$ their ϱ pole and ϱP cut. In addition,

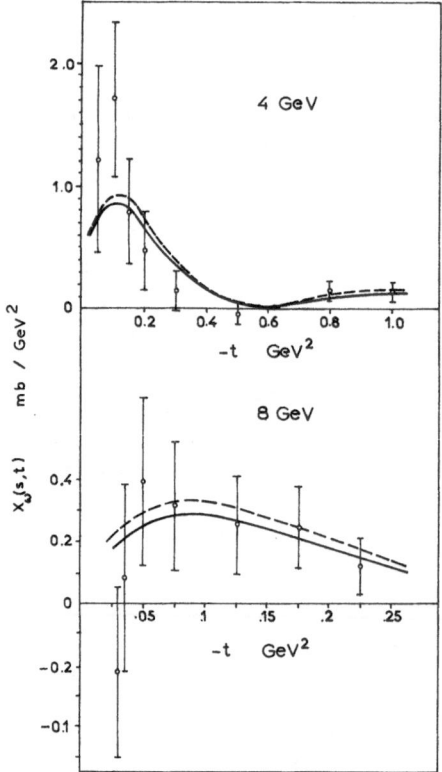

Fig. 11. The ω exchange contribution to $\pi N \to \varrho N$. Data as in Ref. [6]

there is a B-meson Regge pole (denoted by $B = B(s, t)$). It can be shown that [3]

$$M_{\frac{1}{2},\frac{1}{2}} = -(\tfrac{1}{2} A_\varrho + A_{\varrho P} + \tfrac{1}{2} B), \qquad M_{\frac{3}{2},-\frac{1}{2}} \simeq (A_\varrho - B). \qquad (5.8)$$

Thus independently of the magnitude of B:

$$A_\varrho(s, t) + A_{\varrho P}(s, t) = M_{\frac{3}{2},-\frac{1}{2}}(s, t) - M_{\frac{1}{2},\frac{1}{2}}(s, t). \qquad (5.9)$$

The DAM then implies that $\mathrm{Im}(A_\varrho + A_{\varrho P})$ must be a linear combination of $J_2(R\sqrt{-t})$ and $J_0(R\sqrt{-t})$. The imaginary parts of A_ϱ and $A_{\varrho P}$ of the fits are plotted in Figs. 13 and 14; their sum can indeed be well approximated by the proper linear combination of J_2 and J_0. In these fits $\tilde{\lambda}_\varrho$ (nonflip) $\approx 1.9 \sim 2.4$ implying an overall strong cut.

---

[3] In the expression of $M_{\frac{3}{2},-\frac{1}{2}}$ we have neglected the ϱP cut that arises from the transform (2.6) with $n = 2$. This, unless multiplied by a very large $\tilde{\lambda}$ factor, is relatively small. Notice also that the logarithmic factors in our Regge cut contributions have been simplified by replacing $\log(\alpha's) - i\pi/2 \to \log(\alpha's)$ (see Eq. (A) of the earlier footnote 2).

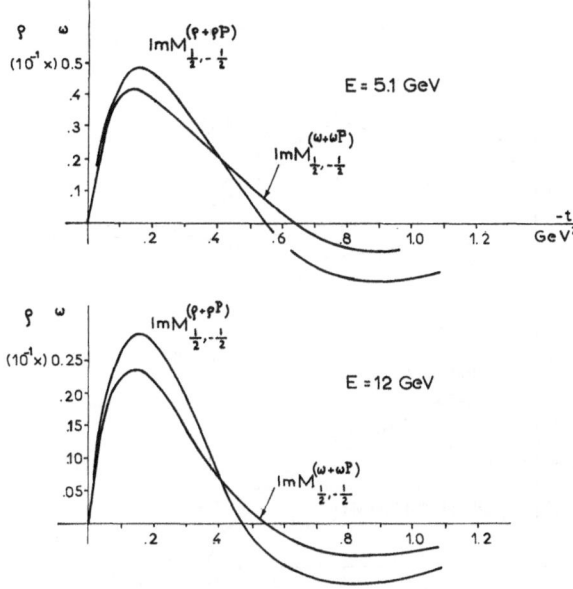

Fig. 12. Helicity flip amplitudes (model B)

(5.C) Nondominant Parts of ϱ and ω Exchange: The helicity flip part of the ϱ-exchange, albeit relatively small, must, according to the DAM, have an imaginary part with the t-structure of $J_1(R\sqrt{-t})$. This is also checked (Fig. 12); in the fits: $\tilde{\lambda}_\varrho$ (flip) ≈ 1. Also, the helicity nonflip and double flip part of the overall ω contribution must have the structure of Eq. (5.9). The check is given in Figs. 13 and 14; now the fits have $\tilde{\lambda}_\omega$ (nonflip) ≈ 2.

(5.D) B-Exchange: The magnitude of B exchange can be determined from the combination of Gottfried-Jackson frame density matrix elements $\varrho_{11}^{(t)} - \varrho_{1-1}^{(t)}$ for $\pi^+ n \to \omega p$ and $\pi^+ p \to \omega \Delta^{++}$. In our model it has the form

$$(\varrho_{11}^{(t)} - \varrho_{1-1}^{(t)})\,d\sigma/dt \sim |B(s,t) + (t + m_\omega^2 - m_\pi^2)\,A_{\varrho P}(s,t)|^2, \qquad (5.10)$$

where $A_{\varrho P} = \varrho$-Pomeron cut (nonzero for both natural and unnatural parity). Hence at $t = -(m_\omega^2 - m_\pi^2) = -0.59\,\text{GeV}^2$ the ϱP cut does not contribute [10] and (5.10) can be used to determine $B(s,t)$. Vector dominance then determines the magnitude of B exchange in photoproduction. The magnitude of B determined in this way is insufficient to eliminate the dip due to the NWSZ of the ϱ-pole; a strong ϱP cut is necessary. This is particularly clear at the highest energies (at 9 GeV for $\gamma p \to \eta^0 p$ and notably at 16 GeV for $d\sigma(\gamma p \to \pi^+ n)/dt - d\sigma(\gamma n \to \pi^- p)/dt$.

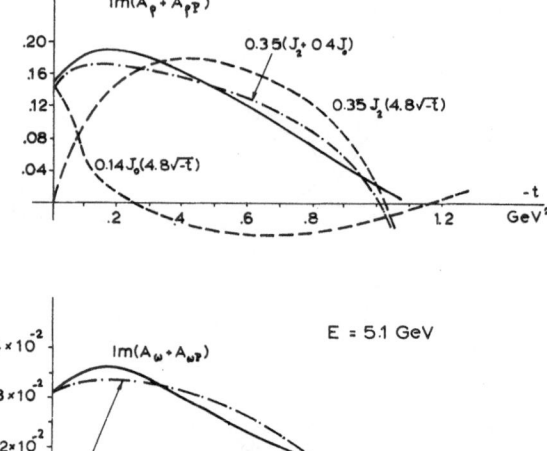

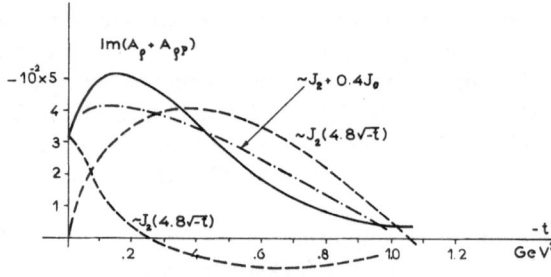

Fig. 13. Helicity nonflip and double-flip amplitudes at 5.1 GeV (model B)

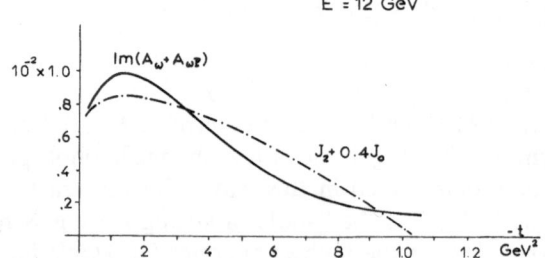

Fig. 14. Helicity nonflip and double-flip amplitudes at 12 GeV (model B)

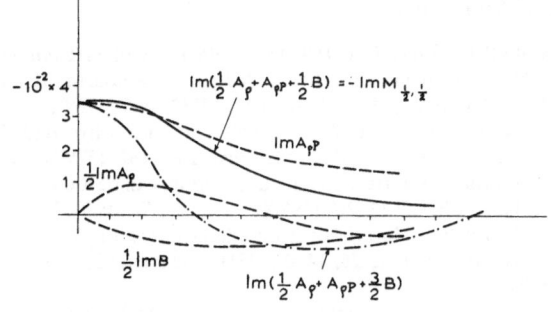

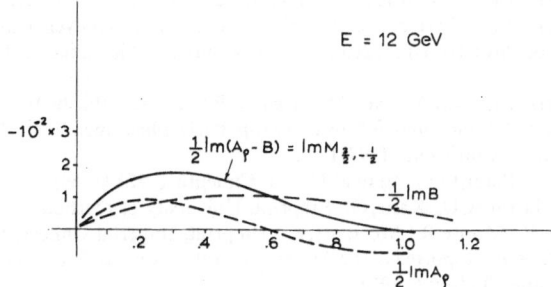

Fig. 15. Nonflip and double-flip amplitude for $\varrho + \varrho P + B$ (model B)

Our conclusion is in agreement with independent analyses of other authors [11]. To account for the 16 GeV data we need a B exchange, violating vector dominance by about one order of magnitude.

In view of the relations (5.8), the DAM demands

$$\mathrm{Im}(\tfrac{1}{2} A_\varrho + A_{\varrho P} + \tfrac{1}{2} B) \sim J_0(R \sqrt{-t}) \quad \mathrm{Im}(A_\varrho - B) \sim J_2(R \sqrt{-t}). \quad (5.11\mathrm{a,b})$$

A plot of the l.h. sides of these relations is given in Fig. 15. The requirement of (5.11b) is seen to be well satisfied; with our B contribution (5.11a) is not so well satisfied. However, a B contribution about 3 times larger satisfies (5.11a) as well (see Fig. 15). Notice that the requirements (5.11a) and (5.11b) also fix the sign of the B relative to the ϱ exchange[4].

Acknowledgements. The author wishes to thank Professors H. Rollnik, K. Dietz and D. Atkinson for helpful discussions and for the hospitality extended to him at Bonn.

---

[4] The recent 5 and 8 GeV data on $\pi^- p \to \pi^0 n$ polarization [12] cannot be explained by exchange of a ϱ plus ϱP cut calculated according to (2.6). To account for them the ϱP cut should either be replaced [13] by exchange of a ϱ' with $\alpha_{\varrho'}(t) \simeq \alpha_\varrho(t) - \tfrac{1}{2}$, or be supplemented by exchange of a ϱ' with $\alpha_{\varrho'}(t) \simeq \alpha_\varrho(t) - 1$. In both cases the imaginary part of the flip and nonflip amplitude can well be constructed in agreement with DAM. In fact, model independent analyses of πp data strongly support the predictions of DAM [14].

References and Footnotes

1. See e.g. Arnold, R. C.: Phys. Rev. **153**, 1523 (1967); – Cohen-Tannoudji, G., Morel, A., Navelet, H.: Nuovo Cimento **48** A, 1075 (1967); – Blackmon, M., Kramer, G., Schilling, K.: Phys. Rev. **183**, 1452 (1969); Nucl. Phys. B**12**, 495 (1969); – Contogouris, A. P., Lebrun, J. P.: Nuovo Cimento **64** A, 627 (1969) and Nucl. Phys. B**13**, 246 (1969).
2. Henyey, F., Kane, G., Pumlin, J., Ross, M.: Phys. Rev. **182**, 1579 (1969).
3. Many of its important ideas (related with absorption) can be found in Ross, M., Henyey, F., Kane, G.: Nucl. Phys. B**23**, 269 (1970); – Dar, A., Watts, T., Weisskopf, V. F.: Nucl. Phys. B**13**, 477 (1969); – Harari, H.: Proceedings of the Liverpool Conference (1969).
4. Harari, H.: Phys. Rev. Letters **26**, 1400 (1971). – Davier, M., Harari, H.: Phys. Letters **35** B, 239 (1971).
5. Barger, V., Cline, D.: Phenomenological theories of high energy scattering. London: Benjamin, W. A. (1969).
6. Contogouris, A. P., Lubatti, H., Tran Thanh Van, J.: Phys. Rev. Letters **19**, 1352 (1967).
7. Van Hove, L.: Phys. Letters **5**, 252 (1963); – Logunov, A.: Nguyen Van Hieu, Todorov, I. T.: Ann. of Phys. (N.Y.) **31**, 203 (1965); – Khuri, N., Kinoshita, T.: Phys. Rev. **137** B, 720 (1965).
8. Contogouris, A. P., Ray, S., Svec, M., Argyres, E.: (To be published).
9. Lovelace, C.: Presence and future of two-particle phenomenology. Invited paper at the Cal. Techn. Conference (1971).
10. Le Bellac, M., Plaut, G.: Lettere al Nuovo Cimento **1**, 721 (1969).
11. Gault, F., Martin, A. D., Kane, G.: Preprint, University of Durham. – Adjei, S., Collins, P., Hartley, B., Moore, R., Morianty, K.: Preprint, Imperial College, London.
12. Bonamy, P., *et al.*: Reported at the Amsterdam International Conference on Elementary Particles, June 30–July 6, 1971.
13. Barger, V.: Phillips, R. J. N.: Phys. Rev. **187**, 2210 (1969).
14. Halzen,, F., Michel, C.: CERN preprint TH. 1355. – Ringland, G., Roy, D. P.: Phys. Letters **36** B, 110 (1971).
15. B-B-F-O Collaboration, Nuovo Cimento **65** A, 637 (1970). – Abrams, G., *et al.*: Phys. Ref. Letters **23**, 673 (1969).
16. Abrams, G., *et al.*: Phys. Rev. Letters **25**, 617 (1970). – A-B-C-Collaboration, Nucl. Phys. B**22**, 428 (1970).
17. Boyarski, A., *et al.*: Phys. Rev. Letters **20**, 300 (1968); **21**, 1767 (1968). – Heide, P., *et al.*: Phys. Rev. Letters **21**, 248 (1968).

Professor Dr. A. P. Contogouris
Department of Physics
McGill University
Montreal, Canada

The Eikonal Model for Regge Cuts in Pion-Nucleon Scattering *

P. D. B. COLLINS and F. D. GAULT

With 11 Figures

Contents

I. Introduction

In the last few years a lot of effort has gone into trying to account for the deficiencies of the Regge pole model for two-body high-energy scattering processes by adding Regge cuts [1]. A variety of prescriptions for calculating these cuts have been suggested (the absorption, K-matrix, eikonal models, etc.) all of which attempt to unitarize the Regge pole, which provides the "potential" or Born term.

The justification for such models has always been in doubt, however. One problem is that these unitarization procedures appear to reproduce (apart from a phase factor) the Amati-Fubini-Stanghellini [2] (AFS) type of cuts, which, at least in perturbation theory, are known to be cancelled by higher-order unitarity corrections [3]. Also the Regge pole itself gives rise to a complex phase-shift, and so must already include at least some of the absorption, and to absorb again probably involves double counting. In one type of absorption model (usually called the Michigan [4] or strong cut model) the dip structures observed in many differential cross sections are accounted for by pole-cut interference, which demands that the cuts have a strength comparable to that of the poles. They must be so strong in fact as to result in an over-absorption of the low partial waves [5], that is to say the positive amplitude obtained from the Born approximation becomes negative due to the enhanced absorption. This defies ones physical intuition as to what absorption is supposed to mean.

---

* Dedicated to Professor G. Höhler on his 50th birthday.

Recently some progress has been reported in justifying the use of the eikonal approximation in perturbation theory [6, 7]. It is well known that the Feynman diagrams which give rise to the leading asymptotic behaviour in two-body scattering are those multiple scattering diagrams which correspond to the break up of the incident particles into component parts, each part scattering on a different component of the other particle. Such diagrams have the non-planar structure which is required to avoid the AFS cancellation, and it is found that, suitably interpreted, the eikonal model does reproduce the sum of such diagrams. If each scattering occurs via the exchange of a Regge pole, these multiple scattering diagrams correspond to Regge cuts.

In the next section we briefly review these results, and go on to show how the model can plausibly be generalized to include any number of different types of Regge exchanges, and how the conventional absorption model is obtained as an approximation. It is also demonstrated that the over-absorption problems of the strong-cut model are automatically cured when the full set of Regge cuts is included.

We then proceed to test different versions of the eikonal model against the πN elastic and charge-exchange scattering data.

It is found that the Michigan model with P, P' and ϱ trajectories gives quite a good account of the cross-sections, but completely the wrong structure for both the elastic and charge-exchange polarizations [8]. The eikonalized "weak-cut" (or Argonne) model, in which the Regge trajectories all choose nonsense, is equally unsatisfactory. We find that a model with strong cuts in the helicity non-flip amplitudes, but no cuts in the flip amplitudes, is best from a phenomenological point of view. In fact when a ϱ' trajectory is also included we can explain most of the πN data, and our amplitudes are in reasonably good agreement with the rather complete analysis of the 6 GeV data obtained recently by Halzen and Michael [9].

We also consider various alternatives in which the P' is replaced by a low-lying ε, as suggested by recent work of the Karlsruhe group [10, 11].

One of the most interesting features of the eikonal model is that it allows one to include the Pomeranchon trajectory with an intercept $\alpha_P(0) > 1$ without the amplitude violating the Froissart bound as $s \to \infty$ [6]. This is because the eikonalization unitarizes the Reggeon Born term, producing cuts which cancel the increasing pole contribution. Nature does not seem to make use of this freedom, however, and any but the smallest deviations from $\alpha_P(0) = 1$ are ruled out by current data.

Some conclusions about the significance of these various types of eikonal-model fit are presented in the final section.

II. The Eikonal Model

Considerable progress has been made recently in verifying that the eikonal expansion approximately reproduces the sum of leading order graphs in quantum field theory [6, 7].

In particular it has been shown that the eikonal model is valid in ϕ^3 theory provided that the momentum of the exchanged propagators is cut off by including a form factor at each vertex. This means that most of the momentum is carried by the external particles which travel up the sides of the diagrams (see Fig. 1), and large momenta are not carried

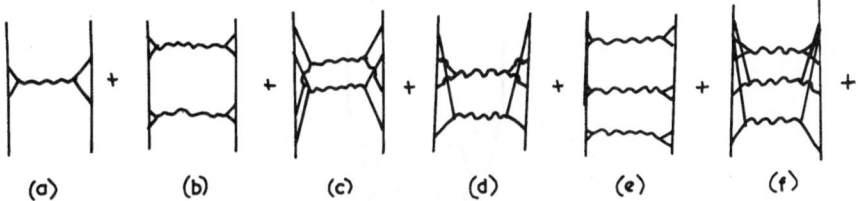

Fig. 1. Multi-Reggeon exchange diagrams. The Reggeon is represented by the wavy line. The eikonal model sums those diagrams in which the couplings have the nested, non-planar structure, i.e. $(a) + (d) + (f) + \cdots$

across the diagram. This seems a very reasonable requirement physically. If the exchanged propagators are replaced by Regge poles (ladders) and the couplings are represented by two-body form factors, as shown in Fig. 1, it is found that the leading order diagrams are those with their couplings "nested" as in Figs 1 d and 1 f. This ensures that each coupling has the third-double-spectral-function "cross" structure which Mandelstam [3] found to be needed if the AFS cancellation, which occurs for planar couplings, is to be avoided.

This cancellation has plagued Regge-cut models ever since Mandelstam's work was published, and it is remarkable that the eikonal prescription can give the correct asymptotic sum of this complicated set of non-planar diagrams using just the on-shell properties of the Regge poles. The physical interpretation of the nested structure is that the incoming particle breaks up into components which scatter separately, and then recombine to form the outgoing particle, as in Fig. 2. Double scattering by one component on another is very improbable at high energies as they do not stay together long enough to exchange more than one Reggeon. Mathematically this is reflected in the fact that planar multiple-scattering diagrams do not contribute to the leading asymptotic behaviour.

If we regard a single Regge pole amplitude $A^P(s, t)$ as the Born approximation, we can calculate the eikonal phase, $\chi^P(s, b)$, that is the phase shift at given impact parameter, by taking the Fourier-Bessel transform [12]

$$\chi^P(s, b) = (1/8\pi s) \int\limits_{-\infty}^{-\infty} dt\, J_0(b\sqrt{-t})\, A^P(s, t).$$ (2.1)

The impact parameter, b, is defined by

$$J \equiv q_s b - \tfrac{1}{2}.$$ (2.2)

J being the total angular momentum, and q_s the centre-of-mass three-momentum. In the eikonal approximation, which should be valid as

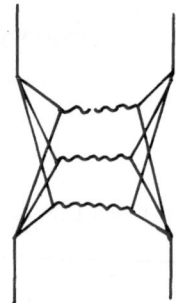

Fig. 2. A re-drawing of Fig. 1 f to show the break up of the incident particles into components which scatter via single Regge pole exchange. All the leading diagrams summed in the eikonal approximation have this structure

$s \to \infty$, $t/s \ll 1$, the impact parameter amplitude is given by

$$\chi^{\mathrm{el}}(s, b) = i[1 - \exp i\chi^P(s, b)]$$ (2.3)

and the full scattering amplitude is the Fourier-Bessel inverse of this

$$A(s, t) = 4\pi s \int\limits_0^\infty b\, db\, i[1 - e^{i\chi^P(s, b)}] J_0(b\sqrt{-t})$$

$$= 4\pi s \int\limits_0^\infty b\, db\, [\chi^P + i(\chi^P)^2/2! - (\chi^P)^3/3! \cdots - i(i\chi^P)^n/n!$$ (2.4)

The first term in this series is the inverse of (2.1), and so gives us back the Regge pole, and the subsequent terms are the cuts due to multi-Reggeon exchange, which we may write in a short-hand notation as $\sum\limits_{n=1}^{\infty} (P)^n$.

There remains the problem that the iteration in t of diagrams like Fig. 1 will result in the renormalization of the "bare" Reggeon input (see Ref. [7]) just as it does in Gribov's Reggeon calculus [13]. At the present time there does not seem to be much that one can do about this, except hope that it will not be too bad an approximation to replace the "bare" Reggeons by the "clothed" physical Reggeons, and neglect these iterations. If this is allowed then the eikonal prescription seems to be justified within the context of perturbation theory, and the extension to strong interactions becomes fairly plausible [7].

In elastic scattering, where we expect the Pomeranchon (P) to be the dominant Reggeon, (2.4) may be used directly if we assume that the P conserves s-channel helicity. More generally spin may be introduced by defining an eikonal phase for each s-channel helicity amplitude [1]

$$\chi_{H_s}^P(s, b) = (1/8\pi s) \int_{-\infty}^{0} dt\, J_{\bar{\mu}}(b\sqrt{-t})\, A_{H_s}^P(s, t) \qquad (2.5)$$

where χ_{H_s} is a matrix in the space of s-channel $(1 + 2 \rightarrow 3 + 4)$ helicities $(H_s \equiv \{\mu_1 \mu_2 \mu_3 \mu_4\})$ and

$$\bar{\mu} \equiv |(\mu_1 - \mu_2) - (\mu_3 - \mu_4)| \qquad (2.6)$$

is the net helicity flip. The full amplitude is then given by

$$A_{H_s}(s, t) = 4\pi s \int_{0}^{\infty} b\, db [\chi^P + i\chi^P \cdot \chi^P/2! - \chi^P \cdot \chi^P \cdot \chi^P/3! \ldots] J_{\mu}(b\sqrt{-t}) \quad (2.7)$$

in analogy with (2.4), where matrix multiplication is implied at each product.

The extension to inelastic processes with quantum number exchange is also fairly obvious. Analogy with the "distorted wave Born approximation" [12] suggests that, if $A_{H_s}^R(s, t)$ is the Born amplitude, given by the Regge pole which carries the quantum numbers, and which has Fourier-Bessel transform $\chi_{H_s}^R(s, b)$, we should put

$$\chi_{H_s}(s, b) = \chi_{H_s}^R(s, b)\, \exp i\chi^P(s, b)$$
$$= \chi_{H_s}^R + i\chi^P \chi_{H_s}^R - (\chi^P)^2 \chi_{H_s}^R/2! + \cdots \qquad (2.8)$$

where the second term is the elastic S-matrix representing multiple Pomeranchon exchange. (We assume that the P conserves s-channel helicity.) If we take only the first two terms of the series (2.8) we reproduce the Sopkovich prescription [14] for absorption in the impact parameter representation. The full series gives us all cuts of the form $\sum_{n=1}^{\infty} R(P)^{n-1}$. The generalization of (2.4) and (2.8) to give any number and type of different Regge exchanges is now obvious. It is only necessary to ensure that the appropriate factorial appears in any term which contains two or more identical Reggeons.

One possible defect of the eikonal method is that we have only included diagrams in which the incoming or outgoing particles appear as *s*-channel intermediate states. It is desirable that one should also take into account all the other different types of diagrams which contain diffractively produced intermediate states having the same quantum numbers as the external particles (such as for example their Regge recurrences) (see Fig. 3). It is obviously impossible to calculate all these diagrams explicitly, but the Michigan group [4], in particular, have made much use of an approximation which includes them by simply multiplying the cuts generated with elastic intermediate states by an over-all

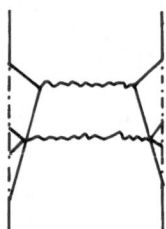

Fig. 3. A diagram to be added to Fig. 1 d showing inelastic intermediate state particles, represented by the dashed lines

enhancement factor $\lambda > 1$. Thus one assumes (rather implausibly) that the enhancement is independent of both s and b. Then instead of (2.8) we have

$$\chi_{H_s}(s, b) = \chi^R(s, b) \exp i\lambda \chi_H^P(s, b) = \chi^R + i\lambda \chi^P \chi_{H_s}^R - \lambda^2 (\chi^P)^2 \chi_{H_s}^R/2! + \cdots \quad (2.9)$$

and correspondingly for elastic processes

$$\chi_{H_s}^{\mathrm{el}}(s, b) = \lambda^{-1} i[1 - \exp i\lambda \chi_{H_s}^P(s, b)]$$
$$= \chi^P + i\lambda (\chi^P)^2/2! - \lambda^2 (\chi^P)^3/3! \ldots . \quad (2.10)$$

One could of course allow different λ's for the different helicity amplitudes and for different Regge exchanges, but we shall deny ourselves this freedom.

The cuts are to be evaluated by inserting the t-channel Regge poles into s-channel helicity amplitudes. Provided the poles have the same nonsense factors, $F(\alpha)$, in all the t-channel amplitudes there is no problem about this. Thus for example at $\alpha = 0$ a nonsense choosing trajectory of even-signature has $F(\alpha) = -1$, while an odd-signature one has $F(\alpha) = i\alpha$, or if a fixed pole is present in its residue $F(\alpha) = i$. We can then write the Regge pole amplitude in the form

$$A_{H_s}^R(s, t) = (-t/s_0)^{\bar{n} + x/2} \exp(-i\pi\alpha(t)/2) \gamma_{H_s}(t) F(\alpha) (s/s_0)^{\alpha(t)} \quad (2.11)$$

γ_{H_s} is the reduced residue in the s-channel amplitude and

$$x \equiv |\mu_1 - \mu_3| + |\mu_2 - \mu_4| - \bar{\mu}. \tag{2.12}$$

With linear trajectories, $\alpha = \alpha_0 + \alpha' t$, and an exponential behaviour for the residue, $G_{H_s} e^{at}$, this is conveniently re-written

$$A_{H_s}(s, t) = (-t/s_0)^{\bar{\mu}+x} \exp(-i\pi\alpha_0/2) F(\alpha) G_{H_s}(s/s_0)^{\alpha_0} \exp ct \tag{2.13}$$

where

$$c \equiv a + \alpha' [\log s - i\pi/2] . \tag{2.14}$$

The Fourier-Bessel transform of such an amplitude is readily found from [15]

$$\int_{-\infty}^{0} \exp(ct) (-t)^{m+\bar{\mu}/2} J_{\bar{\mu}}(b\sqrt{-t}) dt = (b/2)^{\bar{\mu}} (\partial/\partial^c)^m \frac{\exp(-b^2/4c)}{c^{\bar{\mu}+1}} \tag{2.15}$$

and the inverse can be obtained using

$$\int_{0}^{\infty} \exp(-b^2/4c) (b^2)^{m+\bar{\mu}/2} J_{\bar{\mu}}(b\sqrt{-t}) b \, db \tag{2.16}$$

$$= (-t)^{\bar{\mu}/2} (4c^2 \partial/\partial c)^m (2c)^{\bar{\mu}+1} \exp(ct) .$$

Thus for the ϱ pole with $F(\alpha) = i$ we find for the two πN amplitudes $A_{++}(\bar{\mu} = 0)$ and $A_{+-}(\bar{\mu} = 1)$, $x = 0$,

$$\chi^{\varrho}_{++}(s, b) = i \exp(-i\pi\alpha^{\varrho}_0/2) (s/s_0)^{\alpha^{\varrho}_0} G^{\varrho}_{++} \frac{\exp -b^2/4c_{\varrho}}{8\pi s c_{\varrho}}$$

$$\chi^{\varrho}_{+-}(s, b) = i \exp(-i\pi\alpha^{\varrho}_0/2) (s/s_0)^{\alpha^{\varrho}_0} G^{\varrho}_{+-}(b/2c_{\varrho}) \frac{\exp -b^2/4c_{\varrho}}{8\pi s c_{\varrho}} . \tag{2.17}$$

We assume that the P couples only to A_{++}, and χ^P_{++} is similar to the above but with $F(\alpha) = -1$. The multi-Pomeranchon exchange amplitude $\left(\sum_{n=1}^{\infty} (P)^n\right)$ is

$$A^P_{++}(s, t) = \sum_{n=1}^{\infty} iG^P_{++} [\exp -\tfrac{1}{2} i\pi(\alpha^P_0 - 1)] (s/s_0)^{\alpha^P_0}$$

$$\left(-\frac{\lambda G^P_{++} [\exp -\tfrac{1}{2} i\pi(\alpha^P_0 - 1)] (s/s^0)^{\alpha^P_0}}{8\pi c_P s}\right)^{n-1} e^{cPt/n}(nn!)^{-1} . \tag{2.18}$$

If we set $\alpha^P_0 = 1$ we have $G^P_{++} = \sigma^T s_0$, where σ^T is the asymptotic total cross section (from the optical theorem) and this simplifies to

$$A^P_{++}(s, t) = \sum_{n=1}^{\infty} i\sigma^T s(-\lambda\sigma^T/8\pi c_P)^{n-1} e^{cPt/n}(nn!)^{-1} . \tag{2.19}$$

Similarly for the sum of cuts due to the exchange of the rho with many Pomeranchons, $\sum_{n=1}^{\infty} \varrho(P)^{n-1}$, we have

$$A^{\varrho}_{++}(s,t) = i\left[\exp - \tfrac{1}{2} i\pi\alpha_0^{\varrho}\right] (s/s_0)^{\alpha_0^{\varrho}} G^{\varrho}_{++}$$

$$\cdot \sum_{n-1}^{\infty} (-\lambda\sigma^T/8\pi c_P)^{n-1} X_n \exp c_{\varrho} X_n t/(n-1)!$$

$$A^{\varrho}_{+-}(s,t) = i\left[\exp - \tfrac{1}{2} i\pi\alpha_0^{\varrho}\right] (s/s_0)^{\alpha_0^{\varrho}} G^{\varrho}_{+-}\sqrt{-t} \tag{2.20}$$

$$\cdot \sum_{n=1}^{\infty} (-\lambda\sigma^T/8\pi c_P)^{n-1} X_n^2 \exp c_{\varrho} X_n t/(n-1)!$$

where

$$X_n \equiv c_P[(n-1)c_{\varrho} + c_P]^{-1}. \tag{2.21}$$

The first term with $n = 1$ corresponds to the ϱ Regge pole.

The presence of an additional t factor in (2.13) is equivalent to differentiating the amplitudes by c (see (2.15)). Thus if the rho chooses nonsense at $\alpha = 0$ we have $F(\alpha^{\varrho}) = i\alpha^{\varrho} = i(\alpha_0^{\varrho} + \alpha^{\varrho\prime}t)$ for which we apply the operator $\alpha_0^{\varrho}(1 + (\alpha^{\varrho\prime}/\alpha_0^{\varrho}) \partial/\partial c_{\varrho})$ to (2.20) giving

$$A^{\varrho}_{++}(s,t) = i[\exp - \tfrac{1}{2} i\pi\alpha_0^{\varrho}] (s/s_0)^{\alpha_0^{\varrho}} G^{\varrho}_{++} \sum_{n=1}^{\infty} (-\lambda\varrho_T/8\pi c_P)^{n-1}(n-1)!^{-1}$$

$$\cdot \{X_n + (\alpha^{\varrho\prime}/\alpha_0^{\varrho}) [(-(n-1)/c_P) X_n^2 + X_n^3 t]\} \exp c_{\varrho} X_n t$$

$$A^{\varrho}_{+-}(s,t) = i[\exp - \tfrac{1}{2} i\pi\alpha_0^{\varrho}] (s/s_0)^{\alpha_0^{\varrho}} G^{\varrho}_{+-}\sqrt{-t} \sum_{n=1}^{\infty} (-\lambda\sigma_T/8\pi c_P)^{n-1}(n-1)!^{-1}$$

$$\cdot \{X_n^2 + (\alpha^{\varrho\prime}/\alpha_0^{\varrho}) [(-2(n-1)/c_P) X_n^3 + X_n^4 t]\} \exp c_{\varrho} X_n t.$$

Similar formulae may be written for other combinations of pole exchanges, but in this paper we shall confine ourselves to $(P)^n$, $\varrho(P)^{n-1}$, and $P'(P)^{n-1}$, and neglect "Regge-Regge" cuts such as $(\varrho)^n$, $(P')^n$ and $(\varrho)^{n-m}(P')^m$, which are lower-lying J-plane singularities.

This gives us a model for πN scattering requiring 15 parameters; α_0, α' and α, for each of the three trajectories, s_0, σ^T, G^{ϱ}_{++} and G^{ϱ}_{+-}, $G^{P'}_{++}$ and λ. If we take $s_0 = 1$ always, and assume exchange degeneracy of the P' and ϱ trajectories, require that the ϱ trajectory should pass through the ϱ mass, and set $\alpha_P(0) = 1$ this reduces our search to 10 free parameters in both the strong- and weak-cut versions. In later sections we shall alter some of these restrictions, however.

The above amplitudes are to be inserted in the following formulae to obtain the measurable quantities [1]:

total cross-section

$$\sigma^{\mathrm{Tot}}(s) = s^{-1} \operatorname{Im} A_{++}(s,0),$$

differential cross-section

$$d\sigma/dt = (64\pi s q_s^2)^{-1} \{|A_{++}(s,t)|^2 + |A_{+-}(s,t)|^2\},$$

polarization

$$P\, d\sigma/dt = -2\,\mathrm{Im}\{A_{++}(s,t)\, A^*_{+-}(s,t)\}(64\pi s q_s^2)^{-1} \qquad (2.23)$$

The amplitudes for $\pi^{\pm} p$ elastic scattering are $A^P + A^{P'} \mp A^{\varrho}$, and that for $\pi^- p \to \pi^0 n$ is $-\sqrt{2}\, A^{\varrho}$.

III. Models for πN Scattering

In this section we shall compare various Regge pole and cut hypotheses with the πN scattering data.

(a) The Strong Cut Model

Here we assume that $F(\alpha) = -1$ for the P' and $= i$ for the ϱ, so that all the structure of the differential cross-sections as a function of t comes from interference between the poles and the cuts.

Two very important features of the ϱ exchange amplitude that we must reproduce are the cross-over zero at $t \approx -0.1$ where the difference between the elastic differential cross-sections $d\sigma(\pi^- p)/dt - d\sigma(\pi^+ p)/dt$ changes sign [18], and the dip in the $\pi^- p \to \pi^0 n$ data at $t \approx -0.6$. In the old Regge pole fits (see e.g. Ref. [16]) the cross-over zero was obtained by simply inserting a zero in the rho residue. However similar zeros are not found in other processes (see Ref. [17]) and so this explanation is incompatible with factorization.

The strong cut model can certainly account for these facts quite easily, at least in a qualitative way, as pointed out by the Michigan group [4]. For if we take just the first two terms of (2.20), neglect the slopes of the trajectories (which make the c's complex), and put $c_{\varrho} = c_P = 4$, $\sigma^T = 24$ m.b., $s_0 = 1$, we find that A_{++} has a zero at $t = -0.21$, and A_{+-} at $t = -0.55$, as required, if $\lambda = 2$.

But on closer examination this Michigan explanation is very peculiar. The impact parameter Regge pole amplitude for $\bar{\mu} = 0$ is shown in Fig. 4 and we see that it is a maximum for $b = 0$ (i.e. low partial waves from (2.2)). The approximate elastic S-matrix, the first two terms of (2.9), is negative for $b = 0$ and has its peak at the peripheral values $b \approx 1$ fm. Hence on taking the product of the two we find that the low partial waves have been "over-absorbed" [5], i.e. have had their signs changed, which does not make very much sense physically.

The problem occurs only because we have taken the truncated elastic S-matrix, however [7]. This is what the Michigan group [4] do, although

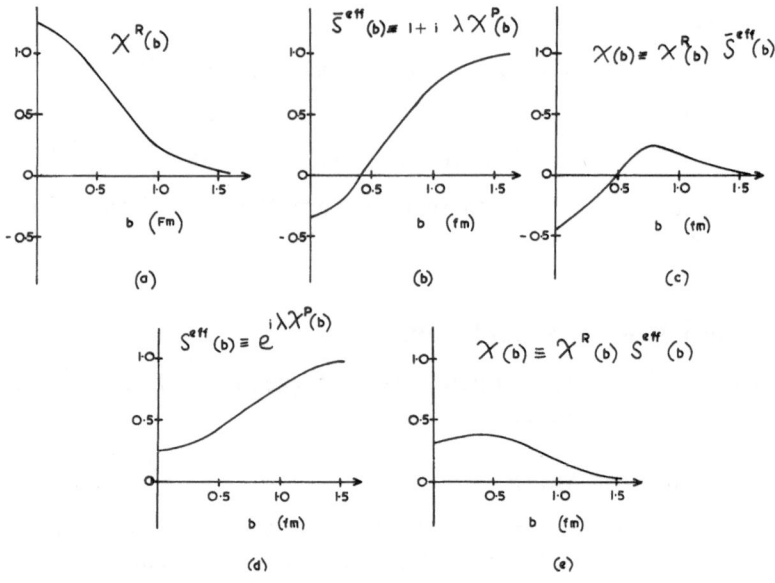

Fig. 4. Impact parameter amplitudes used in the absorption model. a The Regge pole amplitude. b The absorption model approximation to the elastic S-matrix, with enhancement $\lambda = 2$. c The absorbed amplitude showing over absorption for low impact parameters due to the large value of λ. d The full, eikonalized, enhanced, elastic S-matrix. e The absorbed amplitude when eikonalization is used. The parameters used are those quoted in the text

they regard $\chi^P(s, b)$ as the full elastic amplitude, not just the Pomeranchon exchange. If instead we use the full series (2.9), the impact parameter amplitude is reduced, but does not have its sign at $b = 0$ changed by the absorption, as shown in Fig. 4c. In fact only if $\sigma^T \to \infty$ does the absorption become 100%, which makes good physical sense. We thus conclude that the over-absorption problem stems simply from the illegitimate truncation of the eikonal series employed in the conventional absorption model.

Fig. 4 makes it clear that when the full series is used the absorbed amplitude is much less peripheral than it is in the conventional model, which means that in order to get the dips in the required positions a good deal of parameter adjustment is needed. Note that the $\varrho(P)^2$ cut has opposite sign to the ϱP cut, etc. Thus with the same parameters as above it is impossible to get the $\bar\mu = 0$ zero in to $t = -0.2$ however big λ is made. It is essential to vary the c's as well. And we find that $\lambda = 3$ is necessary to get the $\bar\mu = 1$ dip at -0.6. However, since in the strong cut model we expect $\sigma^T(\infty)$ to be greater than it is at current accelerator energies λ can be reduced correspondingly. (Always the product $\lambda \sigma_T$ appears in the cut terms.) We can thus anticipate that the strong cut model with multiple scattering will have somewhat greater difficulty in fitting the data than the conventional Michigan model, and this turns out to be the case.

We have tried to fit the representative set of πN data given in Ref. [19] with P, P' and ϱ trajectories, with P and P' conserving s-channel helicity, no nonsense factors in the poles, and the full eikonal series of $(P)^n$, $P'(P)^{n-1}$ and $\varrho(P)^{n-1}$ cuts. We require that the series be accurate to one part in 10^4 which requires 4–6 terms, depending on the energy, etc. The

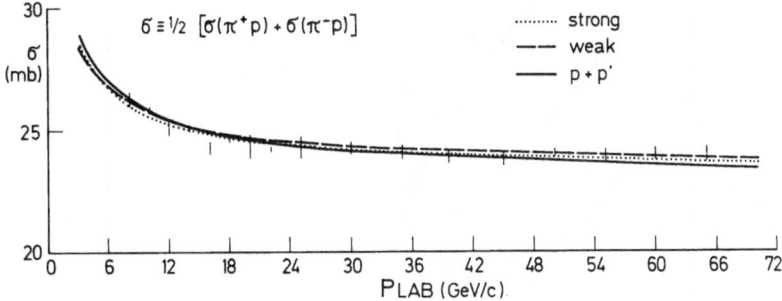

Fig. 5. The best fits obtained with the strong cut model (see Section III (a)), the weak cut model (Section III (b)), and the ϱ with ϱ' model (Section III (d)), compared with the total cross-section data of Ref. [19]

best fit, shown in Figs. 5–9, using the parameters of Table 1, is very similar to that reported by Ter-Martirosyan et al. [20], and gives a good account of the differential cross-sections, including the $t \approx -0.6$ dip in $\pi^- p \to \pi^0 n$. It would appear that the model is consistently a bit below the data for $t < -0.8$, perhaps indicating too great a mutual cancellation between the various cuts, but of course our parameterization is only intended for small $|t|$ so it is hard to assess the severity of this problem. Our total cross-section curve is not completely satisfactory either since the data exhibit too sharp a change of slope at 30 GeV to be compatible with any simple analytic model, but it is quite acceptable given the probable uncertainties in the data.

By far the worst disagreement concerns the elastic and charge exchange exchange polarization data exhibited in Figs. 6c, 7b. The elastic polarizations have two distinctive features, the mirror symmetry of $\pi^\pm p$ about the $P = 0$ axis, and the double zero at $t \approx -0.6$. The symmetry indicates that the polarization is mainly due to interference between the ϱ, which changes sign if $\pi^+ \leftrightarrow \pi^-$, and $P + P'$ which do not. Since P and P' conserve helicity we have, from (2.23)

$$P \, d\sigma/dt \approx -2 \operatorname{Im}[(A^P_{++} + A^{P'}_{++}) A^{\varrho*}_{+-}].$$

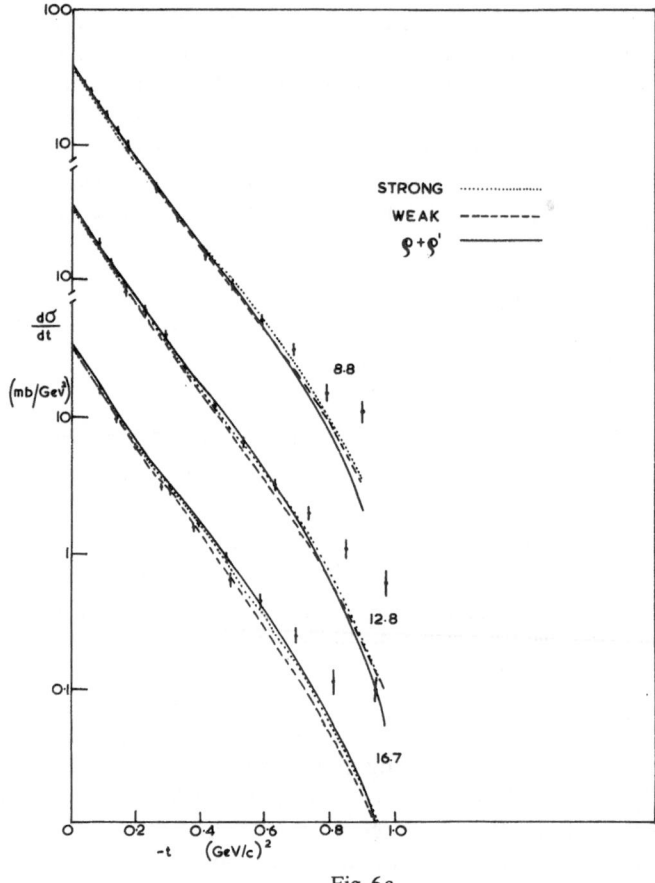

Fig. 6a

Fig. 6. The best fits with the three models (see Fig. 5) compared with the data of Ref. [19] for a $d\sigma(\pi^+ p \to \pi^+ p)/dt$, b $d\sigma(\pi^- p \to \pi^- p)/dt$, and c $\pi^\pm p$ elastic polarization

The sign change which our model predicts at $t \approx -0.4$ is due to the zero of the rho pole amplitude near this point, and is clearly wrong.

Another unsatisfactory feature is that the cross-over zero occurs at $t = -0.3$. Although the precise position of the cross-over is difficult to determine experimentally, because of the uncertainties and incompatibilities in the data near $t = 0$ (especially those associated with the Coulomb corrections), present evidence suggests $t = -0.1$ to -0.15 [9, 18]. We have tried to make a fit in which the cross-over is constrained to occur at $t = -0.1$, but we find that λ has to be increased to ≈ 3, and the description of the data is very poor. Of course increasing λ makes the eikonal series converge more slowly, and distorts the $t \approx -0.6$ dip.

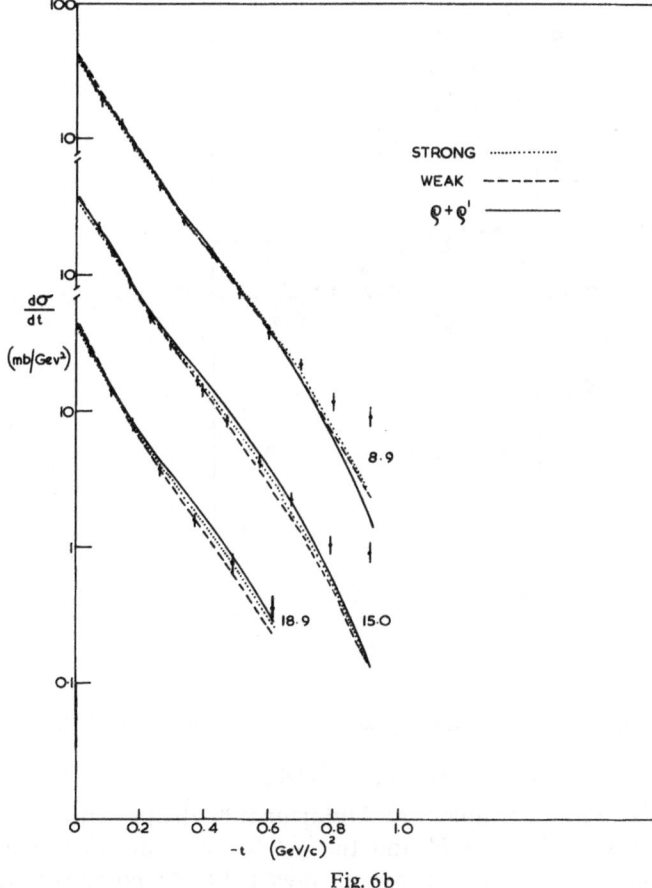

Fig. 6b

b) *The Choosing Nonsense Model*

As an alternative to the above we may allow the rho to choose nonsense, and so put $F(\alpha) = i\alpha$ for both $\bar{\mu} = 0$ and 1 amplitudes. Fits along these lines have been reported in Refs. [21], and in many other earlier works (see Ref. [1]).

Our best fit is reported in Figs. 5–9 and Table 1. Again it is comparatively easy to account for the differential cross sections, but now of course the $t = -0.6$ dip in $\pi^- p \to \pi^0 n$ is due to the nonsense factor of the ϱ. The description of the dip is not as good as model (a) above, however, because the cuts in the helicity flip amplitude are too strong and move the dip in from the point $\alpha = 0$.

The elastic polarization is little better than in (a). This is surprising because the real part of the rho pole amplitude has a double zero at $\alpha = 0$,

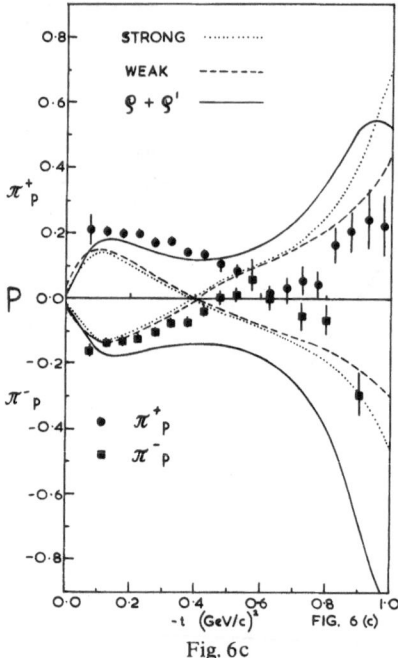

Fig. 6c

and A^P_{++} is almost pure imaginary, so that the main contribution to the polarization is

$$P \approx -2\,\mathrm{Im}(A^P_{++})\,\mathrm{Re}(A^\varrho_{+-}) \tag{3.3}$$

from (3.1), which has a double zero. Unfortunately the presence of other contributions including the P' and the $\varrho(P)^{n-1}$ cuts produces a wide separation of the two zero's, however, indeed the second zero is completely removed in some of our attempted fits. It is clear from Fig. 6c that the data require essentially coincident zeros in the polarization.

There is no significant improvement in the charge-exchange polarization. Another disquietening feature is that the cross-over zero is out at $t = -0.3$. This is because the A^ϱ_{++} pole amplitude has a zero at $t \approx -0.6$, and although the cuts pull it in a bit they are not strong enough to move it in as far as $t = -0.1$. As with the Michigan model we can try to force the zero further in by increasing λ, but we then ruin description of the dip due to the A^ϱ_{+-} amplitude. In any case since the cuts are weaker in the Argonne model the enhancement needed is very great.

c) *The Choosing-sense Model*

To try and improve on this latter aspect we have investigated a model in which the ϱ chooses sense. This means of course that it has different

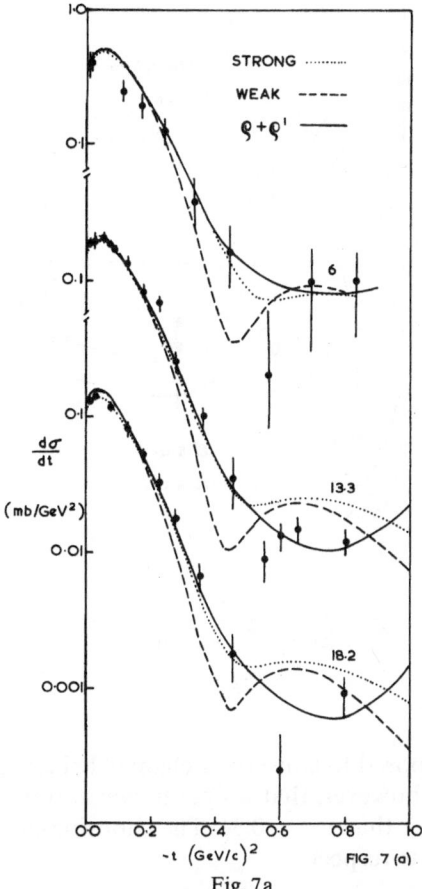

Fig. 7a

Fig. 7. The best fits with the three models (see Fig. 5) compared with the data from Ref. [19] on a $d\sigma(\pi^- p \to \pi^0 n)/dt$ and b charge exchange polarization

α factors in the two t-channel helicity amplitudes, and so it is necessary to perform the crossing explicitly. Since we are only working to first order in t/s in our cut calculations we can approximate the crossing matrix by its $t = 0$ behaviour, and find

$$\begin{aligned} A^s_{++} &= A^t_{++} - (\sqrt{-t}/2m)\,A^t_{+-} \\ A^s_{+-} &= (\sqrt{-t}/2m)\,A^t_{++} + A^t_{+-} \end{aligned} \tag{3.2}$$

where m is the nucleon mass. For the two t-channel ϱ pole amplitudes we take

$$\begin{aligned} A^{\varrho t}_{++} &= i[\exp - \tfrac{1}{2} i\pi\alpha_0]\, G^{\varrho t}_{++} (s/s_0)^{\alpha_0^\varrho} \exp c_\varrho t \\ A^{\varrho t}_{+-} &= (\sqrt{-t}/2m)\, i\alpha [\exp - \tfrac{1}{2} i\pi\alpha_0]\, G^{\varrho t}_{+-} (s/s_0)^{\alpha_0^\varrho} \exp c_\varrho t\,. \end{aligned} \tag{3.3}$$

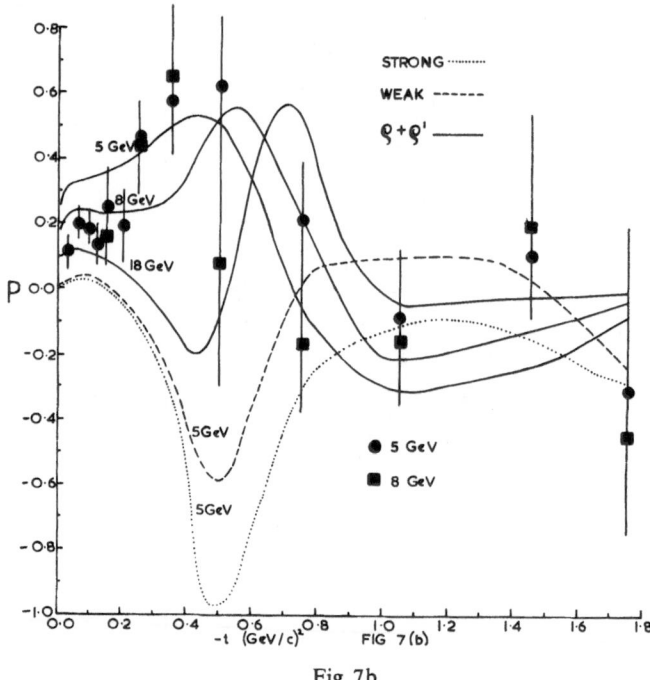

Fig. 7b

The P and P' are supposed to conserve s-channel helicity so no change is needed. It turns out, however, that no parameterization is able to bring the cross-over in closer than $t = -0.36$. The choosing-nonsense model is definitely better in this respect.

d) *No Cuts in the Flip Amplitude*

The problem we noted in section (b), that the approximate double zero in the elastic polarization at $t \approx -0.6$ is obliterated by the cuts and the P', suggests that it might be better to remove the cuts altogether from the s-channel flip amplitude by setting $\lambda_{+-} = 0$, and at the same time give the P' a zero i.e. adopt the non-compensation mechanism for this trajectory. The very successful Regge pole fit of Ref. [16] used this mechanism to get the double zero in the term $\text{Im}(A^{P'}_{++} A^{P}_{+-})$ [see Eq. (3.1) above]. In our case the presence of cuts in A_{++} means that the zeros of the two amplitudes will not be exactly coincident of course.

The lack of success with the cross-over zero in the last section causes us to revert to a nonsense choosing ϱ, but now that there are no cuts in the flip amplitude we have the advantage over section (b) that a large value

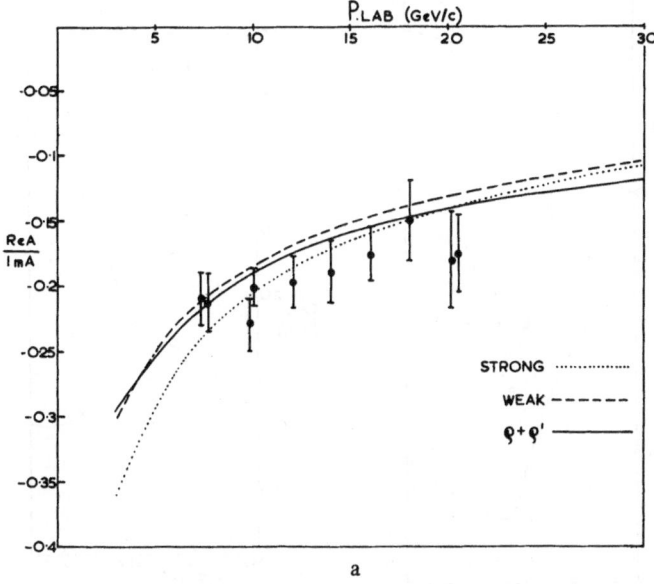

a

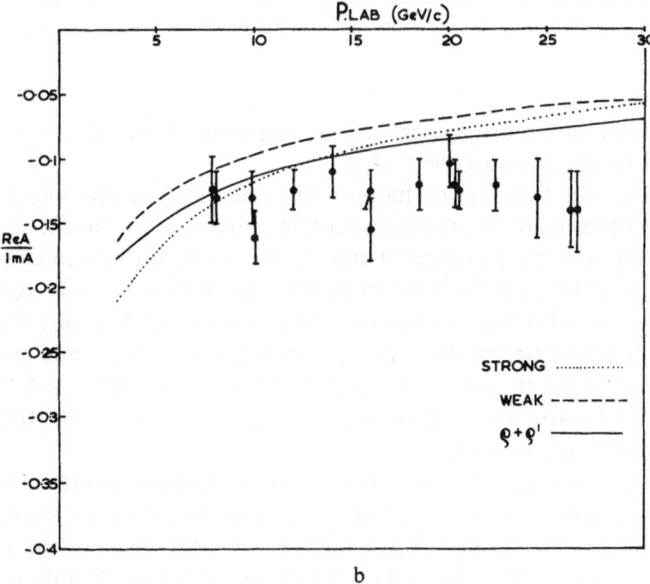

b

Fig. 8. Comparison of the best fits with the three models (see Fig. 5) with the data of Ref. [19] on the ratio of the real to the imaginary part of the forward amplitude for a $\pi^+ p \to \pi^+ p$, and b $\pi^- p \to \pi^- p$

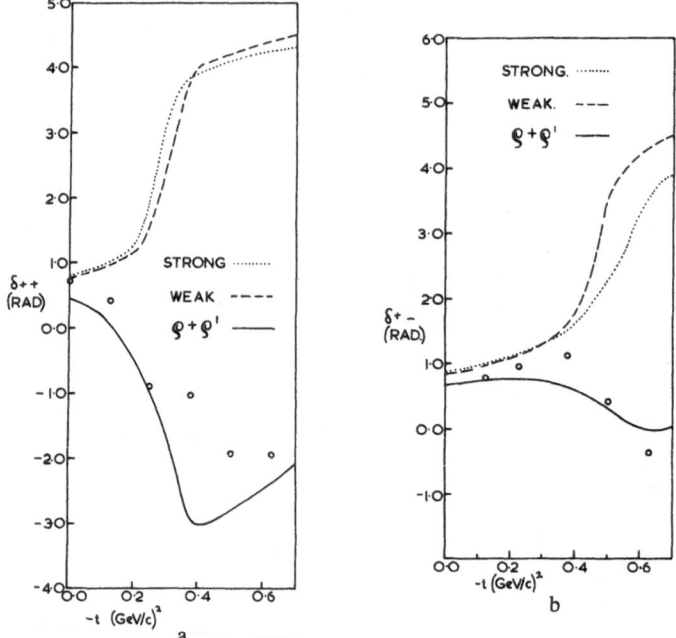

Fig. 9. Comparison of the phases for the $I_t = 1$, s- channel helicity amplitudes for the three models (see Fig. 5) with the "experimental" values obtained by Halzen and Michael, Ref. [9].

of λ will not necessarily spoil the description of the dip at $t = -0.6$, produced by the ϱ nonsense factor.

It is obvious that these changes will not improve the disasterously wrong predictions of the charge exchange polarization that we found in Sections (a) and (b), however. There seems to be no alternative but to follow Ref. [16] and include a secondary ϱ' trajectory in order to try and reproduce the observed behaviour. We assume that it has the same coupling mechanism as the ϱ (i.e. is nonsense choosing) and we fix its trajectory parallel to that of the ϱ with an intercept $\alpha(0) = -0.5$. There are thus 4 extra free parameters G_{++}, G_{+-}, a_{++} and a_{+-} for the ϱ'. We include $\varrho'(P)^n$ cuts as well.

It also proved necessary to allow the slope parameter a to be different for the A^s_{++} and A^s_{+-} rho amplitudes in order to get the cross-over zero in the right place (which needs a large a^ϱ_{++}), and at the same time reproduce the slope of $d\sigma/dt$ at larger t (which depends mainly on a^ϱ_{+-}). Even with this freedom it proved impossible to obtain a dip in the charge-exchange $d\sigma/dt$ at $t = -0.6$ because if one insists on getting a cross-over at $t \approx -0.15$ the $A^{I_t=1}_{++}$ amplitudes has a maximum in this region which

Table 1

Model	Parameters																	χ^2/pt
	P					P'		ϱ						ϱ' or ε					
	α_0	α_1	a^P	G^P_{++}	λ	$G^{P'}_{++}$	$a^{P'}$	α_1	G^{ϱ}_{++}	G^{ϱ}_{+-}	D_{+-}	a^{ϱ}_{++}	a^{ϱ}_{+-}	α_0	G_{++}	G_{+-}	a_{++}	a_{+-}	
Strong	1^a	0.43	0.9	29.4	1.5^a	142	5.2	0.90	25.1	205	—	3.4	$=a_{+-}$	—	—	—	—	—	5.9
Weak	1^a	0.47	1.3	26.4	1.1	101	6.0^b	0.87	42.2	405	—	2.1	$=a_{+-}$	—	—	—	—	—	6.9
ε	1^a	0.64^a	1.7	23.7	0	—	—	0.90	40.0	301	—	.87	$=a_{+-}$	-0.51	-573	—	6.0	$=a_{++}$	8.0
$\varrho+\varrho'$	1^a	0.16	1.7	31.2	1.49	210	3.6	0.80	37.6	319	-0.03	4.6	2.9	-0.50^a	232	281	1.2	0.3	3.0
$\alpha^0_P \neq 1$	1.091	0.14	1.92	20.2	1.48	180	4.8	0.93	25.1	206	—	3.4	$=a_{++}$	—	—	—	—	—	5.9

[a] Fixed parameter
[b] a was bounded between (0,6) for all exchanges.

tends to fill in the effect of the dip of A_{+-}. (We did not find this difficulty with the models of sections (a) and (b) because they do not have the cross-over close enough to $t = 0$). It seems that any model which reproduces the cross-over will inevitably have trouble in this respect unless complicated pole parameterizations are permitted. We ended up by using

$$\gamma^\varrho_{+-}(t) = G^\varrho_{+-}(\exp a_{+-} t + D)$$

but this is not very satisfactory for $|t| > 1.0$.

There are thus 16 free parameters in all.

The resulting fit is exhibited in Figs. 5–9 and Table 1. It is evident that we have succeeded in accounting for most of the data quite well, and in particular the elastic polarization and the phases of the $I_t = 1$ exchange amplitudes are much improved. It should be noted that, in contrast to the pole fit of Ref. [16], by including cuts we have been able to reproduce the Serpukhov data, and have obtained the cross-over zero at $t = -0.16$ (for $s = 12$). The latter is mainly due to a cancellation between the ϱ pole and $\varrho(P)^n$ the cuts, but it obtains some help also from the ϱ' (without which it would be at $t = -0.24$). The cross-over moves out to $t = -0.20$ at $s = 35$.

The large $-t$ elastic data are not very well accounted for because of the onset of an unwanted diffraction minimum at $t \approx -1.2 \, \text{GeV}^2$, though this is beyond the range of data which we have tried to fit. The charge-exchange cross-section at large t is not good either, and it seems clear that a more flexible pole parameterization is needed to explain the amplitudes for $-t > 0.8$. The inclusion of cuts makes things worse if anything at large $|t|$.

It is obviously unsatisfactory to have been forced to include the ϱ' simply in order to "explain" the charge exchange polarization, but at present we know of no more plausible way to account for the data. In Fig. 7b we have included a prediction of the behaviour at higher energies, and we see that the shape is changing rapidly, and approaching the form obtained in Section (b). It is very much to be hoped that we shall soon have polarization data at higher energies so that we can discover whether this rapid energy variation really does occur, and hence whether the polarization is a secondary effect. One price we have to pay for the ϱ' is that we do not get the double zero of the elastic polarizations. The two minima are somewhat separated, though the mirror symmetry is preserved.

In conclusion, although we have managed to fit most of the data using this model it is obvious that a natural explanation of all the important qualitative features has not been achieved. At present there seems to be no model which is able to account for all aspects of the data. A better understanding of cuts is urgently needed.

e) P' or ε?

It has been suggested in recent work by the Karlsruhe group [10] that, in view of the flatness of the Serpukhov total cross-sections, the evidence for the existence of the conventional P' trajectory is very weak. They point out that fits involving P, P' and cuts are simply using the cuts to cancel the energy variation of the P' in the 25–70 GeV range. A thorough discussion of the data has been given in Ref. [10] under various assumptions on the high energy behaviour, but excluding the P' trajectory.

A model along these lines has been published in Ref. [11], where it is proposed that one should invoke only a low-lying secondary $I = 0$ trajectory, the ε, with $\alpha_\varepsilon = -0.5 + t$, whose contribution will thus die away quickly with increasing energy, and should allow the P trajectory to pass through $f(1260)$ by giving it a slope of 0.64 GeV^{-2}. A moderately good pole fit was obtained which we have reproduced. The problem with such a model is that if one calculates the real part of the forward non-flip amplitude by means of the Regge poles only it is positive, because the secondary trajectory has $\alpha(0) < 0$, whereas experimentally it is negative (see Fig. 8) as given by the P' with $\alpha(0) > 0$. (The imaginary part must be positive of course to give a falling σ^{Tot}, so the sign of the coupling is fixed.)

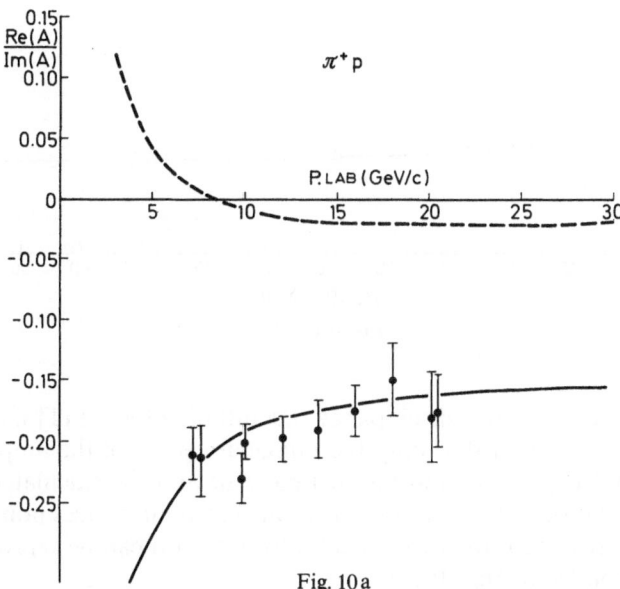

Fig. 10a

Fig. 10. Comparison of the ε model described in Section III (e) with the data of Ref. [19]. Full line: Regge pole model combined with dispersion relations, Ref. [11]. Dashed line: Regge poles and cuts only

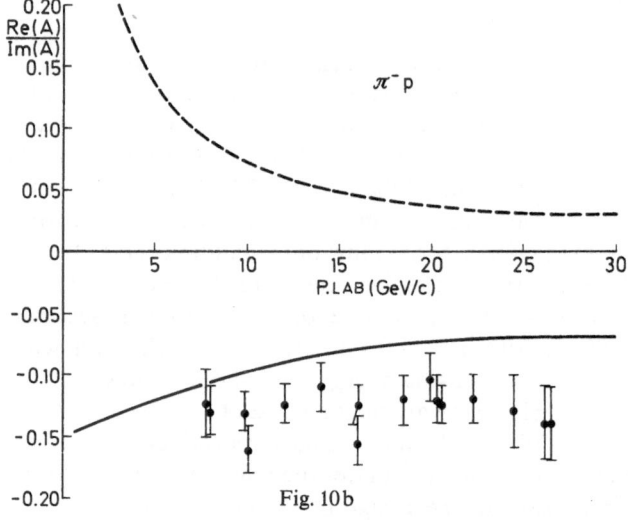

Fig. 10b

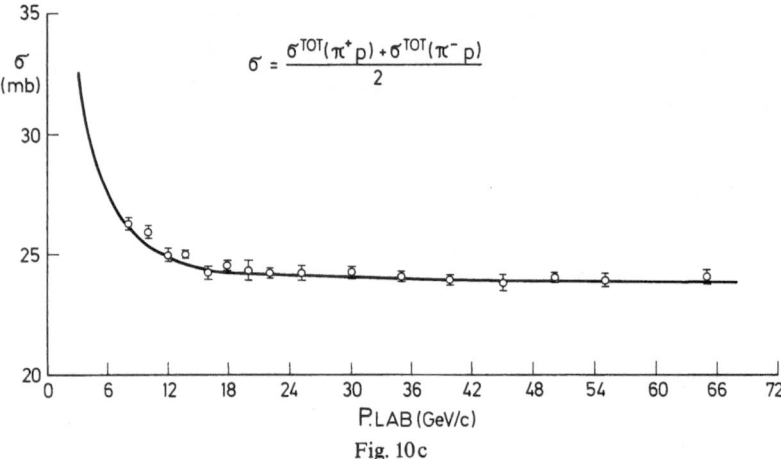

Fig. 10c

In order to avoid this discrepancy the authors of Ref. [11] made the additional assumption that only the imaginary parts of the amplitudes are given by Regge poles and the real parts have to be calculated from dispersion relations. With this prescription, which introduces non-Regge terms in the real part, the experimental Re/Im ratio can be reproduced, as can be seen from Figs. 10a, b.

The introduction of such non-Regge terms seems rather unpalatable, however, and so we have examined the consequences of including cuts as well.

Our model has $(P)^n$, $\varepsilon(P)^{n-1}$ and $\varrho(P)^{n-1}$ cuts. Again quite a good fit to the data on σ, $d\sigma/dt$ is possible, but although the Re/Im ratio is somewhat improved it is still completely wrong, and the elastic polarization is also very poor. We conclude this model is unacceptable unless one is prepared to include non-Regge terms (see Fig. 10).

IV. The Pomeranchon Intercept

The eikonal method described in section II corresponds to a unitarization of the scattering amplitude. This means that even if we start from a Born term with $\alpha(0) > 1$ we shall automatically end up with an amplitude which satisfies the Froissart bound [20] $A(s, 0) \leq \sim s \log^2 s$ because the proof of this bound depends only on the finite range of the forces and unitary. It is thus possible that the Pomeranchon really has an intercept > 1, and that the approximately constant behaviour observed of the total cross sections is in fact due to a complicated supposition of cuts which accumulate at $t = 0$ and cancel the rising pole contribution [21]. The requirement, hitherto maintained, that all trajectories must have $\alpha(0) \leq 1$ was based on the assumption that such an accumulation of cuts would not occur. (Of course a finite number of cuts can never reproduce the asymptotic behaviour of a pole exactly, and so can not effect the necessary cancellation.)

The way the cancellation occurs can be seen in the following way [21]. The impact parameter pole amplitude which gives us (2.18) is

$$\chi^P(s, b) = i \exp[-i\pi(\alpha_0^P - 1)/2] \, (s/s_0)^{\alpha_0^P} \exp(-b^2/4c_P)(8\pi c_P s)^{-1} \quad (4.1)$$

$$\xrightarrow[s \to \infty]{} \frac{i \exp[-i\pi(\alpha_0^P - 1)/2] \, (s/s_0)^{\alpha_0^P}}{8\pi(\alpha^P)' \log(s/s_0)} s^{\alpha_0^P - 1 - b^2[4(\alpha^P)' \log^2(s/s_0)]-1} . \quad (4.2)$$

Hence if

$$b^2 > b_0^2 \equiv 4(\alpha^P)' (\alpha_0^P - 1) [\log(s/s_0)]^2 \quad (4.3)$$

we find $\chi^P(s, b) \xrightarrow[s \to \infty]{} 0$. But if $b^2 < b_0^2$ then $\mathrm{Im} \chi^P(s, b) \to \infty$ and so when we substitute in (2.3), $\chi^{el}(s, b) \to i$.

Hence, since $J_0(0) = 1$, we have, from (2.4)

$$\mathrm{Im} A^P(s, 0) \xrightarrow[s \to \infty]{} 4\pi s \int_0^{b_0} b \, db = 2\pi s b_0^2 . \quad (4.4)$$

So, using the optical theorem (2.23) and (4.3)

$$\sigma^T(S) \xrightarrow[s \to \infty]{} 8\pi\alpha'(\alpha_0 - 1) \log(s/s_0)^2 . \quad (4.5)$$

Thus the Froissart bound is saturated unless $\alpha_0 \leq 1$.

The eikonal model thus predicts that if $\alpha_0^P > 1$ cross-sections will rise $\sim \log^2 s$ at large s, but of course we do not yet have data at high enough energies such that $\log(s/s_0) \gg 1$ to test this. Note that the s_0 which appears in the cut term is the same as that for the poles, i.e. $1\ \text{GeV}^2$. However the eikonal model also makes a very specific prediction as to how this behaviour is approached from below, i.e. we can make use of (2.18) directly to calculate the sum of the cuts at ordinary accelerator energies.

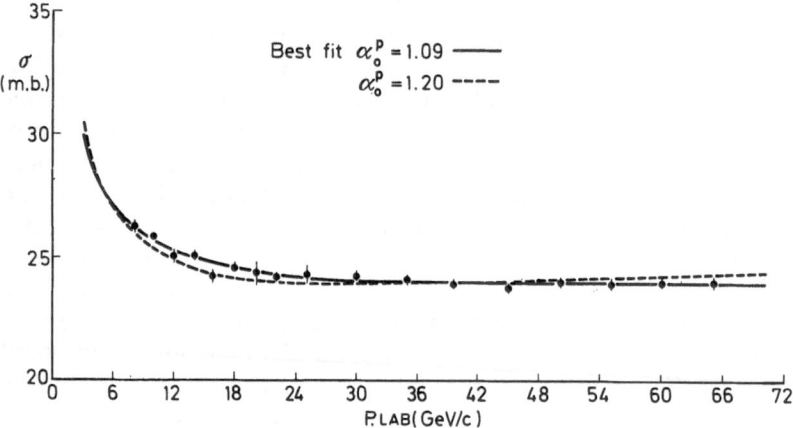

Fig. 11. The total cross sections obtained from best fits to the data of Ref. [19] with a $\alpha_0^P = 1.20$ and b $\alpha_0^P = 1.09$

The best fit with α_0^P fixed at 1.2 is shown in Fig. 11. We see that the model predicts that, once the fall due to the P' is overcome, the total cross-section should begin to rise, because of the rising power behaviour of the P pole, until the cuts become strong enough to cancel the pole, which only happens as $\log s \to \infty$. This is quite contrary to what is observed and so α_0^P can not be much greater than 1. Similarly fits with $\alpha_0^P < 1$ fall with energy contrary to observation. In fact we find that when the intercept is released the best fit has $\alpha_0^P = 1.09$, which is not significantly different from 1.

Of course those who wish to believe that $\alpha_0^P > 1$ are free to regard the eikonal model as valid only for $\log(s/s_0) \gg 1$, and ignore its lower energy predictions. Otherwise it seems that one is compelled to acknowledge the fact that $\alpha_0^P \approx 1$, and that the Pomeranchon saturates (at least to a good approximation), but does not violate, the Froissart bound.

From the viewpoint of the eikonal model this is a pure accident since the eikonalization is quite capable of ensuring that any Reggeon input results in amplitudes which satisfy the bound. Such an accident seems most implausible, in which case one can either take the view that the Pomeranchon is not a Regge pole as we have assumed, and so is not a suitable object to appear as the Born term of an eikonal series, or that the eikonalization procedure is incorrect because the input Regge poles are already unitarized.

By dropping cuts from the flip amplitudes we are already committed implicitly to the view that those poles which couple to $\bar{\mu} \neq 0$ amplitudes can satisfy high energy unitarity approximately by themselves, and now we seem to have been led to the same conclusion for the Pomeranchon, in the non-flip amplitude.

V. Conclusions

We have tested a sequence of Regge pole and cut models, based on the eikonal expansion, against the πN scattering data.

It is found that eikonalization can remove the over-absorption problem of the strong-cut (or Michigan) model, and that both the strong- and weak-cut models can give a reasonably satisfactory account of the differential cross-sections for 3–30 GeV. However it is difficult to obtain good fits to the polarizations, the Serpukhov high-energy total cross section data, and the cross-over zero. The strong-cut model is best for Serpukhov and the cross-over, but hopeless for the elastic and charge-exchange polarizations which are in fact best fitted by poles alone.

The most satisfactory model seems to be one which contains only poles in the spin-flip amplitudes (ϱ and ϱ'), but very strong cuts in the non-flip amplitudes. The inclusion of the ϱ' seems to be the only simple way of reproducing the charge exchange polarization. The final χ^2 is quite good, and the phases of our $I = 1$ amplitudes are in reasonable agreement with those found by Halzen and Michael [9].

We have also considered the possibility that, as suggested in recent work by the Karlsruhe group [11], there is no P' trajectory and that the secondary vacuum pole is the ε with trajectory $\alpha_\varepsilon \approx -0.5 + t$. If the real parts are calculated from dispersion relations the result is satisfactory [11]. However, with just poles this model gives the wrong sign for the forward real part of the non-flip amplitude, and we have investigated whether the inclusion of cuts can help to rectify this discrepancy without the need to include a non-Regge subtraction constant. It is found that even with cuts the model is unacceptable if one wishes to stick to just Regge singularities.

In the eikonal model it is possible for the leading trajectories to have $\alpha(0) > 1$ without the Froissart bound being violated, and so we have allowed the P this freedom. However, unless $\alpha_P(0)$ is very close to 1 it is impossible to reproduce the observed energy dependence of the data.

This suggests that, since the pole alone seems to satisfy the Froissart bound, it must be restricted by, and at least approximately satisfies, s-channel unitarity, and that any model which, like the eikonal model, attempts to unitarize the pole, is guilty of some sort of double counting. The re-normalization problem mentioned in Section II is probably the explanation for this fact. The input to the eikonalization procedure should not be the physical Reggeon but some sort of "bare" Reggeon. Whether the bare Reggeons should satisfy the bound is unclear, but it is probably also a rather academic problem because for phenomenology we need a prescription for calculating cuts from the physical Reggeons.

We conclude that although the eikonal prescription may be on a somewhat stronger theoretical footing than its rival Regge-cut models it is certainly not completely vindicated by the experimental data. The need for just poles in flip amplitudes, but very strong cuts in non-flip amplitudes, stronger even than in the so called strong-cut or "Michigan" type model, seems to be a general feature of high energy scattering, as we discussed at some length in Ref. [7]. It remains therefore (a) to try and understand why cuts seem to be so completely insignificant in flip amplitudes, and (b) to give a more satisfactory method of computing the strong cuts needed in non-flip amplitudes (to explain the cross-over, Serpukhov data, and other points mentioned in Ref. [7]) than the eikonal model.

References

1. Collins, P. D. B.: Physics Reports 1 C, 103 (1971).
2. Amati, D., Fubini, S., Stanghellini, A.: Phys. Letters 1, 29 (1962).
3. Henyey, F., Kane, G. L., Pumplin, J., Ross, M. H.: Phys. Rev. 182, 1579 (1969). — Kelley, R. L., Kane, G. L., Henyey, F.: Phys. Rev. Letters 24, 1511 (1970). — Kane, G. L., Henyey, F., Richards, D. D., Ross, M., Williamson, G.: Phys. Rev. Letters 25, 1519 (1970).
5. Ross, M., Henyey, F. S., Kane, G. L.: Nucl. Phys. B 23, 269 (1970).
6. Arnold, R. C.: Phys. Rev. 153, 1523 (1967). — Togerson, R.: Phys. Rev. 143, 1194 (1966). — Frautschi, S., Margolis, B.: Nuovo Cimento 56 A, 1155 (1968). — Nicoletopoulos, P., Prevost, M. A. L.: Nuovo Cimento 69, 665 (1970). — Tiktopoulos, G., Trieman, S. B.: Phys. Rev. D 2, 805 (1970). — Chang, S. J., Yan, T. M.: Phys. Rev. Letters 25, 1586 (1970). — Hasslacher, B., Sinclair, D. K., Cicuta, G. M., Sugar, R. L.: Phys. Rev. Letters 25, 1591 (1970). — Cicuta, G. M., Sugar, R. L.: Phys. Rev. D 3, 970 (1971). — Cheng, H., Wu, T. T.: Phys. Rev. 186, 1611 (1969). — Cheng, H., Wu, T. T.: DESY report 71/31 (unpublished) and references therein. — Cardy, J. L.: Nucl. Phys. B 28, 455, 477 (1971). — Cardy, J. L.: University of Cambridge report DAMTP 71/2 (unpublished).
7. Collins, P. D. B.: How important are Regge cuts?, Springer Tracts Mod. Phys. 60, 204 (1971).
8. Phillips, R. J. N., Ringland, G. A.: Nucl. Phys. B 32, 131 (1971).
9. Halzen, F., Michael, C.: CERN report Th. 1355 (1971) unpublished.
10. Höhler, G., Steiner, F., Strauss, R.: Z. Physik 233, 430 (1970). — Höhler, G., Strauss, R.: Z. Physik 232, 205 (1970).

11. Achuthan, P., Schlaile, H.-G., Steiner, F.: Nucl. Phys. B **24**, 398 (1970).
12. Glauber, R. J.: In: High energy physics and nuclear structure. Amsterdam: North-Holland 1967.
13. Gribov, V. N.: Soviet Phys. J.E.T.P. **26**, 414 (1968).
14. Sopkovich, N. J.: Nuovo Cimento **26**, 186 (1962). — Jackson, J. D.: Rev. Mod. Phys. **37**, 484 (1965).
15. Magnus, W., Oberhettinger, F.: Functions of mathematical physics, p. 131. New York: Chelsea Publ. Co. 1949.
16. Barger, V., Phillips, R. J. N.: Phys. Rev. **187**, 2210 (1969).
17. Contogouris, A. P., Tran Thanh Van, J.: Phys. Rev. Letters **19**, 1352 (1967).
18. Höhler, G.: Invited talk at Titisee meeting on "Two-body reactions at high energies", 1971 and Proc. of the Sixth Rencontre de Moriond: High energy phenomenology, ed. J. Tran Thanh Van, Orsay (1971).
19. The data used is as follows: σ^{Tot}: Foley, K. J. *et al.*: Phys. Rev. Letters **19**, 330 (1967). — Allaby, J. V. *et al.*, Phys. Letters **30** B, 500 (1969). — $d\sigma(\pi^{\pm}\beta)/dt$: Foley, K. J. *et al.*: Phys. Rev. Letters **11**, 425 (1963). — $P(\pi^{\pm}p)$: Borghini, M. *et al.*: Phys. Letters **24** B, 77 (1967). — $\text{Re} A_{++}/\text{Im} A_{++} \pi^{\pm}p$: Foley, K. J. *et al.*: Phys. Rev. **181**, 1775 (1969). — $d\varrho(\pi^{-}p\to\pi^{0}n)/dt$: Sonderegger, P. *et al.*: Phys. Letters **20**, 75 (1966). — $P(\pi^{-}p\to\pi^{0}n)$: Guisan, O.: Sixth Rencontre de Moriond, see Ref. [18].
20. Boreskov, K. G., Lapidus, A. M., Sukhorukov, S. T., Ter-Martirosyan, K. A.: IETP. (Moscow) preprint, 1971.
21. Arnold, R. C., Blackmon, M. L.: Phys. Rev. **176**, 2082 (1968). — Carreras, B., White, J. N. J.: Nucl. Phys. B **24**, 61 (1970). — Ferber, A. C.: Nuovo Cimento 4 A, 1 (1971). — Drouffe, J. M., Navelet, H.: Nuovo Cimento 2 A, 39 (1971). — Hamer, C. J., Ravendal, F.: Phys. Rev. D **2**, 2687 (1970). — Hong Tuan, R., Kaplan, J. M., Sanguinetti, G.: Nucl. Phys. B **32**, 655 (1971).
22. Froissart, M.: Phys. Rev. **123**, 1053 (1962).
23. See Cheng and Wu, and Cardy, Ref. [6].

SPRINGER-VERLAG
BERLIN·HEIDELBERG·NEW YORK

Springer Tracts in Modern Physics

To appear in forthcoming volumes:

J. Brandmüller and R. Claus: Light Scattering on Optical Phonons and Polaritons
R. Graham: Statistical Theory of Instabilities in Stationary Non-Equilibrium Systems with Applications to Lasers and Non-Linear Optics
D. Schmid: Nuclear Magnetic Double Resonance — Principles and Applications in Solid State Physics

Volume 65
With 94 figures
Approx. 150 pages. 1972
Cloth DM 58,—

H. Theissen: Spectroscopy of Light Nuclei by Low Energy (70 MeV) Inelastic Electron Scattering
H. Arenhövel and H. J. Weber: Nuclear Isobar Configurations
K. Heinloth: Experiments on Electroproduction in High Energy Physics

Volume 64
With 36 figures
III, 100 pages. 1972
Cloth DM 38,—

T. Springer: Quasielastic Neutron Scattering for the Investigation of Diffusive Motions in Solids and Liquids

Volume 63
With 97 figures
V, 189 pages. 1972
Cloth DM 78,—

Photon-Hadron Interactions II: International Summer Institute in Theoretical Physics,DESY, July 12 - 24, 1971
A. P. Contogouris, A. Donnachie, J. Frøyland, F. M. Renard, D. Schildknecht, K. Schilling and P. D. B. Collins, F. D. Gault

SPRINGER-VERLAG
BERLIN · HEIDELBERG · NEW YORK

Springer Tracts in Modern Physics

Volume 62
With 46 figures
V, 147 pages. 1972
Cloth DM 58,—

Photon-Hadron Interactions I: International Summer Institute in Theoretical Physics, DESY, July 12 - 24, 1971
H. D. Dahmen, G. Furlan et al., K. Huang, R. Jackiw, P. V. Landshoff, V. Rittenberg, H. R. Rubinstein, C. H. Llewellyn Smith

Volume 61
With 41 figures
IV, 166 pages. 1972
Cloth DM 68, —

J. L. Basdevant: $\pi\pi$ Theories
A. Donnachie: The Nucleon Resonances
G. Gustafson and J. Hamilton: The Dynamics of Some πN Resonances
B. Schrempp-Otto and F. Schrempp: Are Regge Cuts Still Worthwhile?
R. Oehme: Rising Cross-Sections
B. Renner: On the Problem of the Sigma Terms in Meson-Baryon Scattering
H. Genz: Local Properties of σ-Terms: A Review
P. Brinckmann: Polarization of Recoil Nucleons from Single Pion Photoproduction

Volume 60
With 61 figures
IV, 233 pages. 1971
Cloth DM 78,—

J. Wess: Conformal Invariance and the Energy-Momentum Tensor
R. V. Mendes and Y. Ne'eman: Representations of the Local Current Algebra.
A Constructional Approach
M. Weinstein: Chiral Symmetry. An Approach to the Study of the Strong Interactions
K. Dietz: Dual Quark Models
Chung-I Tan: High Energy Inclusive Processes
J. Drees: Deep Inelastic Electron-Nucleon Scattering
J. J. de Swart, M. M. Nagels, T. A. Rijken and P. A. Verhoeven: Hyperon-Nucleon Interaction
P. D. B. Collins: How Important are Regge Cuts?

Distributor in USA:

Springer-Verlag New York, Inc.
175 Fifth Ave, New York, N. Y. 10010

SPRINGER TRACTS IN MODERN PHYSICS

Ergebnisse
der exakten Natur-
wissenschaften

Volume 63

Reprint

J. Frøyland

High Energy Photoproduction of Pseudoscalar Mesons

Springer-Verlag Berlin Heidelberg New York 1972

SPRINGER-VERLAG
BERLIN·HEIDELBERG·NEW YORK

Springer Tracts in Modern Physics

To appear in forthcoming volumes:

J. Brandmüller and R. Claus: Light Scattering
on Optical Phonons and Polaritons
R. Graham: Statistical Theory of Instabilities
in Stationary Non-Equilibrium Systems with
Applications to Lasers and Non-Linear Optics
D. Schmid: Nuclear Magnetic Double
Resonance — Principles and Applications in Solid
State Physics

Volume 65
With 94 figures
Approx. 150 pages. 1972
Cloth DM 58,—

H. Theissen: Spectroscopy of Light Nuclei
by Low Energy (70 MeV) Inelastic Electron
Scattering
H. Arenhövel and H. J. Weber: Nuclear Isobar
Configurations
K. Heinloth: Experiments on Electroproduction
in High Energy Physics

Volume 64
With 36 figures
III, 100 pages. 1972
Cloth DM 38,—

T. Springer: Quasielastic Neutron Scattering for
the Investigation of Diffusive Motions in Solids
and Liquids

Volume 63
With 97 figures
V, 189 pages. 1972
Cloth DM 78,—

Photon-Hadron Interactions II: International
Summer Institute in Theoretical Physics, DESY,
July 12 - 24, 1971
**A. P. Contogouris, A. Donnachie, J. Frøyland,
F. M. Renard, D. Schildknecht, K. Schilling
and P. D. B. Collins, F. D. Gault**

SPRINGER-VERLAG
BERLIN·HEIDELBERG·NEW YORK

Springer Tracts in Modern Physics

Distributor in USA:

Springer-Verlag New York, Inc.

175 Fifth Ave, New York, N. Y. 10010

SPRINGER TRACTS IN MODERN PHYSICS

Ergebnisse
der exakten Natur-
wissenschaften

Volume 63

Reprint

P. D. B. Collins and F. D. Gault
The Eikonal Model for Regge Cuts in Pion-Nucleon
Scattering

Springer-Verlag Berlin Heidelberg New York 1972

SPRINGER TRACTS
IN MODERN PHYSICS

Ergebnisse
der exakten Natur-
wissenschaften

Volume 63

Reprint

A. P. Contogouris

Regge Analysis and Dual Absorptive Model

Springer-Verlag Berlin Heidelberg New York 1972

SPRINGER TRACTS
IN MODERN PHYSICS

Ergebnisse
der exakten Natur-
wissenschaften

Volume 63

Reprint

A. Donnachie
Exotic Electromagnetic Currents

Springer-Verlag Berlin Heidelberg New York 1972

SPRINGER TRACTS IN MODERN PHYSICS

Ergebnisse
der exakten Natur-
wissenschaften

Volume 63

Reprint

F. M. Renard

ϱ - ω Mixing

Springer-Verlag Berlin Heidelberg New York 1972

SPRINGER TRACTS IN MODERN PHYSICS

Ergebnisse
der exakten Natur-
wissenschaften

Volume 63

Reprint

D. Schildknecht

Vector Meson Dominance, Photo- and Electro-production from Nucleons

Springer-Verlag Berlin Heidelberg New York 1972

SPRINGER TRACTS IN MODERN PHYSICS

Ergebnisse
der exakten Natur-
wissenschaften

Volume 63

Reprint

K. Schilling
Some Aspects of Vector Meson Photoproduction on Protons

Springer-Verlag Berlin Heidelberg New York 1972